PRINCIPLES OF
QUANTITATIVE LIVING
SYSTEMS SCIENCE

International Federation for Systems Research
International Series on Systems Science and Engineering

IFSR was established "to stimulate all activities associated with the scientific study of systems and to coordinate such activities at international level." The aim of this series is to stimulate publication of high-quality monographs and textbooks on various topics of systems science and engineering. This series complements the Federation's other publications.

A Continuation Order Plan is available for this series. A continuation order will bring delivery of each new volume immediately upon publication. Volumes are billed only upon actual shipment. For further information please contact the publisher.

Volumes 1–6 were published by Pergamon Press.

PRINCIPLES OF QUANTITATIVE LIVING SYSTEMS SCIENCE

JAMES R. SIMMS

Simms Industries, Inc.
Fulton, Maryland

Foreword by

James Grier Miller

and

Jessie L. Miller

Springer Science+Business Media, LLC

Library of Congress Cataloging-in-Publication Data

Simms, James R., 1924-
    Principles of quantitative living systems science / James R.
Simms.
    p. cm. -- (International Federation for Systems Research
international series on systems science and engineering ; v. 13)
    Includes bibliographical references (p.   ) and index.

    1. Biological systems. 2. System theory. 3. Animal behavior. I.
Title. II. Series: IFSR international series on systems science and
engineering ; v. 13.
    QH313 .S59 1998
    570'.1'1--ddc21
                                                        98-42066
                                                            CIP

ISBN 978-1-4757-8627-9   ISBN 978-0-306-46966-4 (eBook)
DOI 10.1007/978-0-306-46966-4

© 1999 Springer Science+Business Media New York
Originally published by Kluwer Academic / Plenum Publishers, New York in 1999
Softcover reprint of the hardcover 1st edition 1999
10 9 8 7 6 5 4 3 2 1

A C.I.P. record for this book is available from the Library of Congress.

To my children and grandchildren:
Suzanne, Terry, Patrick, Cassandra, and Kara

Foreword

In 1978, when the book *Living Systems* was published, it contained the prediction that the sciences that were concerned with the biological and social sciences would, in the future, be stated as rigorously as the "hard sciences" that study such nonliving phenomena as temperature, distance, and the interaction of chemical elements. *Principles of Quantitative Living Systems Science,* the first of a planned series of three books, begins an attempt to fulfill that prediction.

The view that living things are similar to other parts of the physical world, differing only in their complexity, was explicitly stated in the early years of the twentieth century by the biologist Ludwig von Bertalanffy. His ideas could not be published until the end of the war in Europe in the 1940s. Von Bertalanffy was strongly opposed to vitalism, the theory current among biologists at the time that life could only be explained by recourse to a "vital principle" or God. He considered living things to be a part of the natural order, "systems" like atoms and molecules and planetary systems. Systems were described as being made up of a number of interrelated and interdependent parts, but because of the interrelations, the total system became more than the sum of those parts.

These ideas led to the development of systems movements, in both Europe and the United States, that included not only biologists but scientists in other fields as well. Systems societies were formed on both continents. In the United States, a group of professors at the Center for the Behavioral Sciences in Palo Alto, California, started a group that, in time and after several name changes, became the International Society for Systems Sciences. At the University of Chicago, a discussion group made up of several of the founding members and others, including von Bertalanffy himself, who spent a year there, met to discuss systems ideas as they applied to the several fields they represented. *Living Systems* included a number of concepts that were developed in those meetings.

The theory described in that book sees living systems as evolving in *levels*—cell, organ, organism, group, organization, community, society, and supranational system—each composed of systems at the level below, organized into twenty subsystems, some of which process matter-energy (matter and energy) and some information. In later work, we added the community between the organization and the society as an additional level. Each subsystem is represented by structural components and carries out specific matter-energy or information processes for the system of which it is a part. Living systems remain in steady states by inputs and outputs of matter, energy, and information from and to the environment.

James Simms approaches the study of living systems from a background in physics and systems engineering and experience in designing systems for the United States military, including one to allow ships to sense the approach of incoming missiles and evade them. This is now in use on U.S. Navy ships. He begins his analysis of living systems with a description of the evolution of the quantitative physical sciences and the measurements used in each and applies this model to develop fundamental concepts and measurements for a quantitative *living* systems science.

Mr. Simms' book, although it is based on living systems theory, is not a restatement of that theory, nor is it simply an application of its concepts. Instead it extends living systems theory by analyzing the fundamental role of energy in the behavior of living things. In his theory, living systems can behave only to the extent that they have available energy which they are able to direct. Therefore, a specific behavior can be quantified by the energy it uses. The capacity to direct energy is a fundamental behavioral characteristic of any living system and is a function of the system's structure and organization. By searching the available literature in the life sciences, he demonstrates that measurements are available that make it possible to quantify behavior of many kinds in both plant and animal organisms.

Information, however, is not neglected in Simms' work, since behavior is determined by one of several sorts of information. A muscle, for example, twitches only when neural information is transmitted to it. Communication from another individual in the form of sensory information can cause coordinated muscle contraction like running away from a threat. He shows that information, like energy, can be measured and expressed in an equation that embodies the system's capacity to direct energy, the amount of available energy, and information.

It is our opinion that this book represents an important step in the development of a quantitative living systems science. As Simms shows, the concepts of *available energy* and the *capacity to direct energy*, as well as the causative relationship between information and behavior, are useful in the analysis of behavior. The systems with which this first book of the series is concerned are mainly at the level of the cell and the animal organ and organism. They include such systems as neurons, motor units, the leg muscle of *Rana pipiens*, and the hearts, respiratory organs, and digestive tracts of various species. It will be interesting to see how the science is applied in later volumes to the more complex behavior of human beings, groups, and higher-level systems.

James Grier Miller
Jessie L. Miller
La Jolla, California

Preface

Principles of Quantitative Living Systems Science is the culmination of over three decades of research by the author. The concept for this research was formulated while I worked as a systems engineer on the Titan International Continental Ballistic Missile (ICBM). My job included analysis of the accuracy of the Titan ICBM for the purpose of improving its destructive capability. It was readily apparent that technology provided a means of destroying civilization. The thought of a war being fought with ICBMs had a major impact on me—especially after having been involved in amphibious operations in the Pacific during World War II and seeing firsthand the horrors of war.

It was easy to rationalize working on ICBMs by understanding that technical people designing and making weapons do not cause wars or start wars—political leaders do! However, this knowledge does nothing to solve the problem of the potential destruction of civilization. Being trained in the quantitative physical sciences and knowing that our civilization is based, in large part, on the technical innovations of man, I thought the solution to the destruction problem lay in the development of a quantitative political science. This science was to have analytical and predictive properties equivalent to those of the quantitative (hard) sciences. This political science was to bring discipline to social innovation in the way that the physical sciences bring discipline to technical innovation. In my naiveté, I thought the task would not be too difficult and could be finished within just a few years. I did not realize then that the physical sciences have evolved over many thousands of years!

A major hypothesis of my research is that fundamental measures and principles can be discovered for a quantitative political science which are equivalent to those of the physical sciences. This hypothesis has proven to be true. The approach to proving the hypothesis was to determine the way the physical sciences evolved and the characteristics associated with the fundamental principles of these sciences. Thus, the evolution and characteristics of the physical (nonliving systems) sciences provided a model for the development of the fundamental principles for a quantitative living systems science.

Inasmuch as the physical sciences are based on measures of the fundamental parameters of the science and relationships among these parameters, the research was started by looking for fundamental measures in the extant political science. None were found. Then the characteristics of the extant social sciences (the super set of political science) were analyzed for fundamental measures. None existed. The behavioral sciences (the super set of the social sciences) were analyzed for possible fundamental measures, with only limited success. The research was then expanded to include all living things (living systems) in an attempt to identify fundamental parameters, their measures, and the relationships among the parameters. Here, too, success was limited. However, at the living systems level, I was able to identify the fundamental parameters, the extant measures, and the missing measures. It was now possible to develop the principles of a living systems science.

The fundamental parameters of living systems behavior were discovered to be: (1) the system's behavior (measured by the energy in the behavior), (2) a unique characteristic of each system—its capacity to direct energy, (3) the energy available to a system, and (4) the behavioral information that causes a living system's behavior. I discovered the nature of behavioral information and established units of measure for behavioral information. Methods were also devised for measuring a system's capacity to direct energy. Discovery of the fundamental parameters and their measures provided the basis for establishing quantitative relationships among these parameters. These parameters and the relationships among them allowed the development of the principles of living systems science.

The fundamental principles resulting from the research apply to the behaviors of all living systems. These fundamental principles for living systems are equivalent to the fundamental principles for the quantitative physical sciences. These principles provide the basis for the development of quantitative living systems, behavioral, and political sciences.

The original research objective, to develop a political science with quantitative precision and predictive capabilities equivalent to those of the nonliving systems sciences, has yet to be achieved. However, the principles upon which such a political science can be based have been developed.

This book is the first in a series of three. It describes the evolution of the quantitative physical sciences, identifies the fundamental parameters of the living systems science, identifies the measures associated with these parameters, and develops the relationships among these parameters as they apply to individual animals. A second book will further define the fundamental parameters and measures as they apply to group behaviors of animals and will develop quantitative relationships for group behaviors. A third book will treat the principles as they apply to (1) the human animal, (2) the evolution of the human species, (3) the technical and social innovations of humans, and (4) quantitative social and political sciences.

Over the years, a number of people have reviewed and commented on my research and made much-appreciated improvements in my efforts. They include

Fred Adler, Bruce Baird, Sue Compton, Raymond Hoop, Patsy Jackson, Suzanne Johnson, Ram K. Khatri, Ida Mae Lundy, Pauline Poe, Charles Simms, Thomas Simms, and Rita Walljasper. Dawn Frick, Sabrina Knouse, and Dave West have also helped with the preparation of this book.

Contents

CHAPTER 1

Introduction

This book presents the development of the principles of quantitative living systems science and the application of these principles to individual animals, including the human species.

This chapter will (1) define special words and terms, (2) provide background on the development of quantitative living systems science, (3) state the fundamental principles of quantitative living systems science that have been developed, (4) describe the methods used to develop these fundamental principles, (5) describe the approach that was used to develop the fundamental principles of quantitative living systems science, (6) describe the models used to guide the development of the fundamental principles, and (7) explain the organization of this book.

The term *living systems* has evolved during the past three decades to differentiate living things from nonliving things. Living systems science applies to all living things and includes the more specialized biological, behavioral, and political sciences. The term *quantitative living systems science* is used here to signify a science based on objectively measurable phenomena. The *natural sciences* include physics, chemistry, and biology. The term *scientific method* is used in its usual meaning, that is, principles and procedures for the systematic pursuit of knowledge involving the recognition and formulation of a problem, the collection of data through observation and experimentation, and the formulation and testing of hypotheses.

The successes of the quantitative natural sciences are generally understood. It is also generally accepted that the social sciences lag the natural sciences in precision and quantification. The differences between the natural and social sciences in the general public's mind is typified by the often-asked question: Why is it that we can send men to the moon but that the human condition remains so bad? In other words, why can't the behavioral and social sciences have precise and quantitative characteristics similar to those of the natural sciences?

The great achievements in the natural sciences during World War II and in the second half of the twentieth century led to a renewed interest in applying natural science methods and techniques to the "soft" sciences (behavioral, social, and political sciences). The efforts to make living systems science more quanti-

1

tative include (1) increased use of the scientific method, (2) use of the physical science concepts of cybernetics and information theory, (3) applying system science methods and techniques, and (4) searching for similarities (isomorphs) between the natural sciences and a living systems science. For example, Shannon's (1949) classical information theory has been applied to the social sciences, but with no measurable success. Other efforts have broadened the application of the natural sciences to all things that live (living systems), but these, too, have not been successful.

Other applications of natural science methods and procedures to the living systems sciences have been successful. Miller's (1978) *Living Systems,* which classifies living systems, is one example. Robert Rosen, in his book *Fundamentals of Measurement and Representation of Natural Systems* (1978) relates the observation and measurement of behavior to energy. Howard Odum, awarded the Craaford Prize for his contribution to biology, has generated a significant body of work with his application of energy concepts to the environment and ecology. Despite these isolated successes in making living systems science more like the physical sciences a formal set of principles of quantitative living systems has not been developed previously.

The method used to develop the principles of quantitative living systems science was based on the hypothesis that the concepts and methods used to develop the physical sciences could serve as a model. This hypothesis requires an understanding of the physical sciences and the way these sciences evolved, and testing it requires the delineation of the evolution and the characteristics of the physical sciences.

The author's research over the past three decades resulted in development of the fundamental principles of quantitative living systems science. These principles are:

1. The behaviors of living systems are observable and measurable by way of the energies used in these behaviors.
2. Living systems have a unique behavioral characteristic which is a capacity to direct energy.
3. A system's capacity to direct energy is a function of its structure and organization.
4. A system's capacity to direct energy can be quantified (measured or calculated).
5. A system's behavior is a function of the energy available to the system.
6. A living system's behavior is a function of behavioral information.
7. Behavioral information is the ability to cause work and can be measured by the work it causes.

8. A living system's behavior is a direct function of the system's capacity to direct energy, to the energy available to the system, and to behavioral information.

The fundamental principles of quantitative living systems science were developed using the scientific method, the methods and procedures used to develop the principles of the existing quantitative sciences, and the methods by which the extant quantitative sciences evolved. These methods involve (1) observing the characteristics of the subjects of the science, (2) identifying the fundamental determinants of these characteristics, (3) establishing measures for these fundamental determinants, and (4) determining the functional relationships among the fundamental determinants, as follows:

- Identify the methods and procedures that were used in the evolution of the quantitative sciences.
- Develop models based on the evolution of the quantitative sciences that can be applied to quantitative living systems science.
- Identify the subject of quantitative living systems science—a system's behavior—in a way that allows measurement.
- Verify that a system's behavior can be quantified (measured or calculated).
- Identify a system's characteristics that provide it with a capability to behave.
- Verify that a system's behavioral characteristics can be quantified.
- Demonstrate that a capacity to direct energy is a system's fundamental behavioral characteristic.
- Using the models, identify the other fundamental parameter of living systems behaviors—behavioral information.
- Determine the nature of behavioral information.
- Devise methods for measuring behavioral information.
- Establish units of measure for behavioral information.
- Formalize the relationships between living systems behavior and the determinants of this behavior.

Evolution of the extant quantitative natural sciences depends not only on development of the fundamental principles of these sciences but also on acceptance of these principles by the user community. A quantitative natural science will be accepted only if the current qualitative living systems science paradigm is changed to quantitative one. This involves (1) identifying the methods and procedures that were used to establish the extant quantitative sciences, (2) developing a model based on these extant established sciences that could be applied to quantitative living systems science, (3) using the model to establish, through test and validation procedures, the principles of quantitative living systems science, (4) validating the principles for the autonomous, volitional and nonvolitional behaviors of animals, and (5) demonstrating how the principles apply to the total behaviors of individual animals.

1.1. Model for Emergence of the Quantitative Sciences

The methods and procedures for the evolution of the quantitative sciences were identified by analyzing the emergence of these sciences. The emergence of the extant quantitative sciences provided a model for the development of quantitative living systems science. The development of this model is presented here.

1.1.1. Quantitative Science Characteristics

The quantitative physical (nonliving systems) sciences were developed using the scientific method. Emergence of the quantitative sciences required knowledge of the phenomena pertaining to the subjects of the science and of the precise relations among these phenomena based on observations and experimentation concerning the quantities involved. The observations and experimentation must result in evidence of a quantitative form for it to have definite meaning. Evidence of this type is obtained by *measurement*, one of the most important elements in all scientific work. Observation, experimentation, and measurement form the basis for the scientific method.

The successes of the quantitative nonliving systems sciences, such as the mechanical, heat, and electrical sciences, are often cited as models for the behavioral and social sciences to emulate. These models were, in fact, used to develop the principles for quantitative living systems science. In order to develop the principles for quantitative living systems science, the quantitative concepts and methods of the nonliving systems sciences must be extended to the living systems sciences. The key elements of this approach are:

- Identify the fundamental characteristics of the quantitative sciences.
- Show how the quantitative sciences evolved as a function of fundamental measures and existing quantitative sciences.
- Derive the unique fundamental measures for quantitative living systems science.
- Relate these unique fundamental measures to existing quantitative sciences.
- Show the development of the basic equations of living systems behavior.

1.1.2. Evolution of the Quantitative Sciences

The evolution of each of the quantitative physical sciences is based on (1) an understanding of the phenomena concerning the behaviors of the subjects of the science, (2) the discovery and/or use of one or more fundamental measures, and (3) the quantitative sciences which existed before the evolution of the new quantitative science.

Fundamental measures are essential to the development of quantitative sciences. Fundamental measures have distinct characteristics:

- The concept associated with the measure is self-evident, that is, axiomatic. For example, the concept of length is self-evident because we have two eyes which allow depth perception. Also, it takes more effort for us to walk to a more distant place than to one that is near.
- The measurement unit is defined in terms of an invariant, or approximately invariant, physical phenomenon.
- The measurement unit is arbitrary, that is, it must be accepted by a consensus of the users of the unit.
- The fundamental measure cannot be reduced.
- The mathematical concept of counting is understood so that one unit, that is, a single count, can be determined.

Figure 1.1 illustrates the evolution of the physical sciences and provides a convenient means for summarizing the essential elements of this evolution. The evolution of fundamental measures is shown from left to right across the lower portion of Figure 1.1. The boxes show fundamental laws and quantitative sciences, and the lines between the boxes show the fundamental measure and the extant quantitative science necessary for the evolution of the next quantitative science.

Length, time, and mass, as shown in Figure 1.1, are fundamental measures of the quantitative sciences. Each has the characteristics of fundamental measures listed previously. The concept of length is axiomatic (self-evident) because we have visual depth perception and it requires more effort to walk to a more distant place than one that is near. The concept of time is also self-evident since we can easily observe the aging process in ourselves, and in the living things around us. The concept of mass is axiomatic as it takes more effort to lift a larger mass than a smaller one. Petrie (1951) has shown that there have been many different units of length in the history of humankind. One such unit is the cubit, which was used in Egypt from the time of the predynastic period onward. In Europe, prior to the nineteenth century, each country had its own system(s) of weights and measures. In China, units differed in value from place to place, and in the same locality people in different trades had conflicting measurement units. Various units of time and mass, have also been traced to ancient times. For example, the oldest stan-

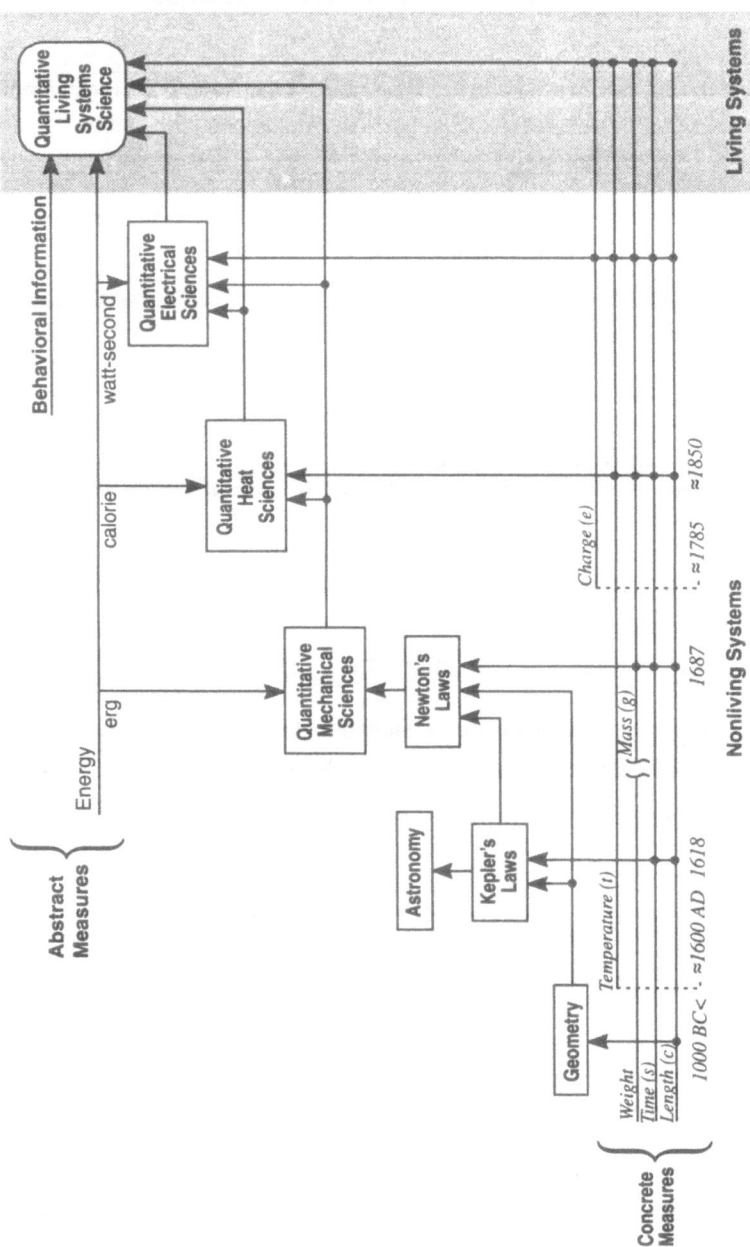

Figure 1.1. Emergence of the quantitative sciences.

dard of mass known is the beqa, found in early Amratian graves in Egypt (7000–8000 B.C.). Currently, the English system, established at the beginning of the nineteenth century, and the metric system, established in 1791, are the widely accepted systems of fundamental units.

Geometry is shown in Figure 1.1 as the first quantitative science. The generally accepted account (Newman, 1956) of the origin and early development of geometry is that the ancient Egyptians needed to invent it in order to restore the landmarks which were destroyed by the periodic inundations of the Nile. The very name *geometry* is derived from two Greek words meaning measurement of the earth. These measurements are based on the fundamental measure of length. Since geometry is, in essence, the relationship among various lengths, no other fundamental measure is necessary for this quantitative science.

Although not included in Figure 1.1, Copernicus provided an understanding of the basic phenomena of astronomy when he conceived the heliocentric theory—reviving Pythagorean beliefs—in his famous book *De Revolutionibus Orbium Coelestrium*. Johann Kepler used the fundamental measures of distance and time and the quantitative science of geometry to bring quantitative precision to Copernicus' theories (Newman, 1956). Kepler used these measures to develop quantitative relationships for the motion of planets. These relationships, known as Kepler's laws, provide the fundamental principles for the quantitative science of astronomy.

The unit of time also meets all the requirements of a fundamental unit. It is self-evident that time passes; the unit of time is based on the invariant physical phenomenon of planetary motion, and the definition of a time unit is arbitrary and is agreed upon by the community of users of time measures. As shown in Figure 1.1, the fundamental measures of length and time, along with the measurement science of geometry, are all prerequisites for the development of the quantitative science of astronomy.

Kepler's work resulted in three fundamental laws. The first states that the planets move in ellipses with the sun at one focus. The second law states that the line joining sun and planet (the radius vector) sweeps out equal areas in equal times. The third law connected the times and distances of the planets: "The square of the time of revolution of each planet is proportional to the cube of its mean distance from the sun."

The next quantitative science to evolve was Newton's mechanics, as shown in Figure 1.1. Newton investigated those phenomena associated with mass, particularly the attraction of masses to one another. These investigations required a fundamental measurement unit of mass. Newton's laws were the result of investigations into the relationships among length, time, and mass. They depended on the preexistence of fundamental measurement units of length, time, and mass, and on the existence of geometry and Kepler's work. Newton's work is of universal significance and changed the current of human thought.

Newton's *Principia*, consisting of three books, records the basis for his mechanics. The first book treats his laws of motion and lays down his mechanical foundations, clearly formulated for the first time. The second book is devoted to motion in a resisting medium and is the first treatment of the motion of real fluids. The third book of the *Principia*, published in 1687, is the crown of the work and changed the face of science. Newton opens the third book with a recapitulation of previous themes and a short statement of the new theme. For the new theme he states, "It remains, that from the same principles, I now demonstrate the frame of the System of the World." In the third book he establishes the movements of the satellites around their planets and of the planets around the sun on the basis of universal gravitation.

The next phase of evolution, as shown in Figure 1.1, is the quantitative science of heat. The nature of heat and its relationship to temperature had been debated at least since the time of the ancient Greeks. It was not until the invention of the thermometer that a fundamental measure of temperature was established and a quantitative heat science could evolve.

Because our knowledge of the properties of matter can be gained only through sense perception, every fundamental quantity must be ultimately defined in terms of perceived sensations. The temperature of a substance may thus be qualitatively defined as the property of the substance that determines the sensation of warmth or coldness received from contact with it (a self-evident truth). Observation of the substance reveals that if a change in temperature is taking place, there is a simultaneous change in other physical properties of the substance. These changes may include volume, pressure (at constant volume), electrical resistance, radiation from the surface, and a change in state. These changes in the various physical properties are usually susceptible to accurate measurement.

The earliest devices used to measure temperature were known as thermoscopes. The invention of the thermoscope has been variously ascribed (Cork, 1942) to Galileo Galilei in 1593, C. Drebbel in 1609, Paolo Sarpi in 1610, Sancttario Santario in 1611, and to G. della Porta in 1610.

The measure of temperature and the self-evident warmth and coldness prove that temperature is a fundamental measure. That is, it is self-evident because we feel the warmth or coolness of substances, the measure is related to measurable physical phenomena, the unit of measure is arbitrary and is determined by the users of the measure, and the measure cannot be reduced to a more fundamental measure.

The quantitative science of electricity is shown in Figure 1.1 as the next step in the evolution of the quantitative sciences. This science could not evolve until the fundamental measure of charge was discovered/invented. A fundamental measure of charge was established by Coulomb in his electrical papers, which were published in the *Memoirs de l'Academie Royale des Science* between 1785 and 1789. The charge on the electron was established by Robert A. Millikan (1935) in 1913. Both these measures are based on the mechanical sciences. The coulomb is based on the force of repulsion between like charges. The charge on the electron is based on attraction of unlike charges. Force is a concept of the mechanical sciences. The conversion of electrical energy to mechanical and heat energies is directly coupled to the quantitative mechanical and heat sciences.

In addition to the fundamental measures based on the above-described self-evident truths, Figure 1.1 shows a measure of energy. Unlike the measures of length, time, mass, temperature, and charge, which are based on concrete matter, the measure of energy is based on an abstract concept. That is, energy is weightless and does not occupy space. The fundamental definition of energy is the ability or capacity to do work. Energy is measured by the amount of work performed. The work performed can be mechanical, thermal (heat), or electrical. Therefore, the types of energy include mechanical, electrical, and heat. The work associated with the various forms of energy has been identified and measurement units established, for example, the erg for mechanical energy, the calorie for heat energy, and the watt-second for electrical energy. Work is a dynamic phenomenon because it is associated with time, such as the watt-second, and thereby can be considered a behavior. Because of these differences in characteristics between energy and the fundamental measures of length, time, mass, temperature, and charge, energy has been identified as an abstract concept in Figure 1.1.

Figure 1.1 shows the evolution of another quantitative science, namely quantitative living systems science, and the evolution of another fundamental measure, namely behavioral information. The fundamental measure of this science has been identified, and a method for measuring behavioral information developed (Simms, 1991, 1996). The quantitative living systems science and the behavioral information concepts are relatively new, and are explained more thoroughly in later chapters.

Analysis of the evolution of the quantitative sciences yields the following results:

- Development of the quantitative sciences is truly evolutionary, with each new quantitative science depending on the sciences that preceded it.
- Measurement of the phenomena associated with the subjects of the science is essential to the development of a quantitative science.
- Each of the quantitative sciences evolved only after the establishment of the fundamental measure of a basic parameter associated with that science.
- Fundamental measures and units of measure relating to the science are prerequisites to the development of a quantitative science.

These facts provide the basis for a model for the development of quantitative living systems science.

1.2. Prerequisites for Quantitative Living Systems Science

One prerequisite for the development of a quantitative science is the ability to measure the phenomena associated with the subjects of the science. A primary observable phenomenon of the subjects of the quantitative sciences is the way they behave. Therefore, it is necessary to be able to measure the behaviors of living systems, just as it is necessary to measure the behaviors of the subjects of the other quantitative sciences. The process used for the measurement of behaviors is described below with particular emphasis on the behaviors of animals. The primary measure of observable behavior is the energy utilized in the behavior, which is discussed further in Chapter 2.

Another prerequisite for the development of quantitative living systems science is the identification of the determinants of the behaviors in terms of the characteristics of the subjects of the science. The structure and organization of living systems are shown in Chapter 3 to be fundamental determinants of the behaviors of these systems. These determinants are typified and measured in terms of the system's capacity to direct energy. Another fundamental determinant of a living system's behavior is the energy available to the system which can be used in the behavior.

Another determinant of a living system's behavior is behavioral information. The dependence of behavior on behavioral information can be illustrated by considering the most easily observable behavior (motion) of animals at two levels. At the level of an individual animal, a command (behavioral information) from one animal to another animal to perform some action can cause the commanded animal to perform the action (behavior). That is, there is a direct relationship between the behavior and the behavioral information. The observed behavior of the commanded animal is the direct result of muscle contractions. At the second level, the level of the muscle, the muscle will only contract if there is an impulse in the motoneuron connected to the muscle tissue. The impulse that activates the muscle contraction is the result of the information input to the commanded animal. It is proven in Chapter 4 that observable animal behavior can only occur if (a) there is behavioral information and (b) this information caused the contraction of the muscle tissue which results in observable behavior.

If quantitative living system science is to evolve, the hypothesized model requires that the new fundamental determinant of behavior (i.e., behavioral information) have a measure. A measure for behavioral information is developed in Chapter 4.

A physical science model for the development of a quantitative science also requires functional relationships among the parameters of a science. Such functional relationships for quantitative living systems science are developed in Chapter 5. Quantitative relationships among these fundamental parameters are also developed in Chapter 5.

1.3. Fundamental Concepts

The model of the evolution of the quantitative sciences and the prerequisites for these sciences as described above were used to develop the fundamental concepts of quantitative living systems science. Like the quantitative science model and the prerequisites, development of the fundamental concepts evolved with a number of false starts and blind alleys. It took a long time to understand that the behaviors of systems could be measured by the energy content in these behaviors. It was another thing to prove conclusively that measures for energy could be used to measure behaviors.

The fact that behaviors of systems can be measured by the energy contents in these behaviors provided a means for determining one of the fundamental determinants of behavior. The heat sciences provided the conservation of energy law. This law states that energy is neither created nor destroyed. Therefore, the energy that caused an observed behavior had to be available from some source. Since energy can be measured, it was apparent that the available energy determinant of behavior can be quantified.

It is readily observable that the behaviors of systems are a function of the structure and organization of the system. For example, the skeletal and muscular structures of different animals result in their motion behaviors being dissimilar. These characteristics of systems have been extensively studied and analyzed and there is a robust literature on these characteristics. It was apparent that the energy in a behavior was different from the available energy that was converted to the behavioral energy. It also became apparent that these differences were the result of the system operating on the available energies in a way that was determined by the characteristics of the system. For example, the differences in the motion behaviors of dissimilar muscles is a function of the structure of the muscle and the skeletal structure to which it is attached, even though the available chemical energy in the dissimilar muscles is the same. An extensive analysis of the relationships among the energies in behaviors, available energy, and the structures and organizations of systems resulted in the identification of a characteristic that is a fundamental determinant of behavior. This determinant is a system's *capacity to direct energy*. This determinant, as shown in subsequent chapters, can be quantified.

Investigation of the readily observed behaviors of animals, from the very small to the largest, revealed that these behaviors are the result of muscle tissue contraction. Muscle tissue contraction takes place if, and only if, there is activation of this tissue. Therefore, activation of muscle tissue is a fundamental determinant of behavior. The form of this activation was investigated extensively and is described in Chapter 4. However, this fundamental determinant does not have a measure. As stated previously, the model for the development of quantitative science requires that a new fundamental determinant have a measure.

1.4. First Principles

The premise of this book is that the processes associated with the evolution of the quantitative nonliving sciences are applicable to the evolution of quantitative living systems science. Evolution of the quantitative nonliving systems sciences proceeds from identification of fundamental principles and measures to the continuing development of these sciences. Research on the development of quantitative living systems science is in the first stage of evolution: establishing the fundamental principles and measures.

The first principles for quantitative living systems science are based on the readily observable behaviors of animals:

1. Behaviors can be observed by way of the energies utilized in these behaviors.
2. Behaviors can be measured.
3. Behaviors are a function of a living system's capacity to direct energy.
4. A system's capacity to direct energy is a function of its structure.
5. Behavioral information is a fundamental measure.
6. A system's behavior is a direct function of its capacity to direct energy, its available energy, and its behavioral information.

The next step is to use these first principles to develop quantitative living systems science.

1.5. Organization of This Book

An overview of the first five chapters was given above. The remaining chapters further detail the first principles and develop quantitative living systems science as they apply to the behaviors of individual animals.

The first step was to develop a model for the evolution of a quantitative science that could be used in the development of quantitative living systems science. Such a model, based on the evolution of the quantitative physical sciences, is developed in Chapter 6. This model includes considerations of how new ideas are accepted and contribute to the evolution of a science. These considerations are based on the work of Kuhn (1970) in his book, *The Structure of Scientific Revolution*. This model also includes a determination of the evolution of the qualitative living systems science to date in order to provide a basis for its further evolution. The status of living systems science evolution is discussed in Chapter 6 and provides a basis for the further evolution of living systems science.

Development of quantitative living systems science requires the validation of the fundamental principles described in the first five chapters. The evolution of the science model presented in Chapter 6 was used as a guide to validate these fundamental principles. This model indicates that validation should begin with the application of fundamental principles and concepts to the most readily observable and ubiquitous living systems. These systems are the ones related to autonomous animal behaviors. Autonomous behaviors are those caused by information generated internally in an animal, such as heartbeat, which is caused by the heart's pacemaker. Testing and validation of the fundamental principles using the most obvious autonomous animal behaviors, that is, heartbeat and blood circulation, breathing, and digestion, are described in Chapter 7, which also validates the fundamental principles as they apply to those autonomous behaviors associated with neural information.

The autonomous behaviors discussed in Chapter 7 provide the basis for the development of another unit of measure for behavioral information in addition to those developed in Chapter 4. This unit of measure is developed in the Appendix. It is based on the heart muscle as the elemental behaving system as opposed to a single skeletal muscle motor unit behaving system. The energy utilized by the heart, in response to a single information input, is large compared to the energy utilized by a single motor unit. This larger unit of measure is more appropriate for measuring the more obvious behaviors such as heartbeat, breathing, and digestion.

After the principles for quantitative living systems science were validated for autonomous animal behaviors, they were applied to the nonvolitional behaviors of animals. These nonvolitional behaviors are additional autonomous behaviors that are based on information generated as a function of environmental phenomena. Nonvolitional behaviors include those known as stimuli–response behaviors. The principles were tested for the behaviors caused by information generated by an animal's sensors in response to thermal, gravitational, light, and mechanical phenomena in an animal's environment. These behaviors, over which animals have no control, include responses to various environmental phenomena. For example, we shiver and perspire in response to changing thermal environments and cannot stop or modify these behaviors. Application of the principles to nonvolitional behaviors and the testing and validation of the principles are treated in Chapter 8.

Testing and validation of the principles as they apply to volitional behaviors are treated in Chapter 9. Volitional behavior is used herein as those behaviors caused by two or more information generators that combine in a neuron and result in an "adjustment" of behavior. For example, a single, deeper-than-normal breath caused by the autonomous information of breathing combining with information from our brain is a volitional behavior. The testing and validation of the principles for volitional type behaviors were performed using food acquisition and thermal behaviors. The principles were validated for these behaviors.

The principles as applied to the total behavior of individuals are presented in Chapter 10. The total behavior of individuals is a combination of the behaviors caused by genetic, chemical, and neural information. The principles were applied to single cell animals, simple multicellular animals, coelenterates, vertebrates, and humans. Tests were performed using existing data and the principles were validated for total behaviors.

Chapter 11 is a summary of the research and the major findings of this research.

CHAPTER 2

Quantification of Behavior

2.1. Introduction

Two of the most important elements of the model presented in the last chapter for the development of a quantitative science are the characteristics of the subjects of the science and the quantification of the behaviors of these subjects. As pointed out in Chapter 1, a prerequisite for the development of quantitative living systems science is the ability to quantify the behaviors of these systems when they are subjected to various conditions. The methods for quantifying the behaviors of living systems are developed in this chapter.

The subjects of quantitative living systems science are all things that live. These systems have been previously classified by the taxonomy of Carolus Linnaeus—often referred to as the "father of classification." A further classification has been made by James G. Miller (1978) in his book *Living Systems*. Miller classifies living systems into a hierarchy of eight levels of increasing complexity ranging from the lowest level of the cell to the highest level of a supranational system. Each of Miller's levels is composed of twenty critical subsystems. Both Linnaeus' and Miller's classification systems are descriptive as opposed to quantitative. However, they do fulfill part of the requirement of our model. They describe the structural characteristics of living systems. Miller also identifies and describes processes associated with living systems.

In addition to the classification of living systems by structural features, there is a robust literature describing in detail the structures and processes of a large number of plants and animals. However, the literature is not adequate for the quantification of the behaviors of these plants and animals. The emergence model requires that the behaviors of the subjects of a science be quantitative as opposed to descriptive.

Before proceeding with the development of quantitative living systems science, it is necessary to describe the characteristics of the nonliving systems emergence model. This description of the model is needed to provide insight into how a living systems science should evolve and to provide a basis for more detailed comparison between the nonliving and the living systems sciences.

As indicated in Chapter 1, a first step in the evolution of a science is to identify the subject of the science. The subject of geometry may be described as the properties of space. Geometry evolved over a long period of time and can be considered in the following stages: ancient empirical geometry, ancient demonstrative geometry, sporadic development during medieval times and the Renaissance, analytical geometry, modern synthetic geometry, and, finally, the foundations of geometry. The early empirical geometry, probably invented in ancient Egypt, consisted merely of a number of crude rules for the mensuration of various simple geometric figures. The ancient Greeks developed these crude beginnings into the science of demonstrative geometry, namely, plane and solid geometry.

In the mid-1600s, René Descartes invented analytical geometry. Analytical geometry is distinguished from the older or synthetic geometry by its method rather than by its content. From the point of view of our model, the stages of the evolution of geometry up to the invention of analytical geometry were, to a large extent, descriptive science. Analytical geometry added the concept of variables— for example, a curve in the plane is represented by an equation in the variable coordinates (x,y) of a point on the curve. The concept of variability includes the principle of behavior. For example, how does a property of space act (i.e., behave) when one of the variables of space changes? Descartes' invention combined the powerful methods of algebra with geometry to form analytical geometry, thereby providing a method for relating the behavior of the properties of space as a function of the determinants of these properties. The technical innovation of analytical geometry broadened the subject of geometry to include the behavioral properties of space. Analytical geometry provided methods for quantifying the behaviors of the properties of space as a function of the determinants of behavior. As indicated in our model, the science of geometry continues to evolve through synthetic geometry and the foundations of geometry.

The subject of astronomy is the celestial bodies. Astronomy includes the magnitudes, motions (behaviors), and constitution of the celestial bodies. The evolution of astronomy followed stages similar to those for geometry. The ancient Egyptians, through their observations of celestial bodies and their concepts of geometry, identified and described the positions of many of the celestial bodies. They named many celestial bodies and constellations of these bodies. They could also describe and predict many behaviors (motions) of celestial bodies such as the moon and the sun. The Greek theories of astronomy were also descriptive, as were Copernicus' heliocentric theories of astronomy. However, they could not quantify the behaviors of celestial bodies, such as their distances from the earth at various times and their speeds. The invention of the telescope and the innovations of Kepler provided the methods for quantifying the behaviors of planets in terms of the quantitative parameters of these behaviors.

The subjects of mechanics are matter and mechanical energy and their interactions. The empirical stage of evolution of the mechanical sciences can be traced back to early man through man's artifacts, such as Stone Age implements. These implements were matter (stone) shaped into forms that could be used as tools for obtaining food and for protecting themselves. These tools provided man with mechanical advantages he did not have prior to these mechanical innovations. Although the artifacts provide a record of the implements of early man, the behaviors of man in the use of these implements left no direct records but had to be deduced or inferred. From these early innovations up to Newton's time, the subjects of the mechanical sciences could be described by the characteristics of mechanical things and the behaviors of mechanical things could be described to a certain extent. However, quantitative measures of behavior were not available until Newton's innovations were developed.

The heat and electrical sciences followed the same evolution patterns as geometry, astronomy, and the mechanical sciences. In the beginning, they also were only descriptive sciences. It was not until the inventions of quantitative measures that behaviors of the subjects of heat and electrical sciences could be measured and quantified.

Our nonliving systems model requires that the behaviors of the subjects of the science be quantified prior to the development of a quantitative science.

The current stage of evolution of the living systems sciences is the end of the empirical stage. A large body of literature exists about living systems, with the preponderance of this literature being descriptive. Transition from the empirical stage of evolution to the quantitative analytical stage depends on an ability to quantify the behaviors of living systems.

One early finding (Simms, 1983) was that the behavior of systems can be quantified in terms of the energies utilized in these behaviors. This chapter provides the development of methods for the quantification of behaviors of systems.

2.2. Definition of Behavior

Behavior, in its most general meaning, is the way things act. In this context, behavior includes the acts of both living and nonliving systems and is used by astronomers, physicists, chemists, biologists, psychologists, behaviorists, and social scientists to describe the observed activities of the subjects of their particular sciences.

The methods for observing behavior are described below along with the measurable behavioral phenomena associated with the observed behavior. All behavior, as will be demonstrated, can be measured, and these measurements can be expressed in the common units of the energies used to perform the behaviors. This measure of behavior provides the foundation for the identification of a causal relationship between behavior and its basic determinants.

2.3. Observation of Behavioral Phenomena

Observation is fundamental to all sciences because the scientific method begins with observation. The basic business of science is to ascertain relationships between observables. A science becomes quantitative, that is, a hard science, when the observables can be measured and the relationships between observables are quantitative. The principles of scientific observation are well established. For example, in his book *Fundamentals of Measurement and Representation of Natural Systems*, Rosen (1978) demonstrates that observation is based on the energies associated with the system being observed. A short review of the major methods for observing behavior provides a background for quantifying behavior.

2.3.1. Methods of Observation

Behaviors can be observed by a number of methods, the most prevalent being our visual senses. However, all five senses of human observers can, to some degree, be used to observe behavior.

The visual senses provide a means for observing (1) obvious behaviors of individuals, such as walking, physical work, and eating, and (2) the more subtle behaviors, such as hand and finger movements associated with writing, sewing, and eating. Most of these descriptive behaviors are manifestations of motion and can be better described in terms of the more fundamental muscle tissue contraction motions. These behaviors are the result of muscle activities of the individuals. These muscle activities are caused by the contractions of the tissue making up the muscles. Since a major characteristic of all animals is their rapid motion, which is caused by contractile tissue, our visual sensors allow us to observe the obvious behaviors of animals.

Although much less important, an observer's sense of hearing can be used in the observation of some behaviors such as the sounds associated with body functions. The heartbeat is an example of an audible observable behavior.

Behaviors can be observed, to some degree, by the sense of touch. For example, one of the most important behaviors of all warm-blooded animals is the metabolic process which generates heat. This can be observed by the sense of touch. The sense of touch also provides observers with the capability to determine the presence of mechanical pressure in a frequency range from zero to the low audio frequencies. An individual's pulse is a mechanical (pressure) behavior observable by the sense of touch.

The sense of smell can be used to observe a very limited range of behaviors. It can be used to observe the scents given off by individuals and used in conjunction with other sensors to observe the behavior of individuals in response to specific scents. Although most people can recognize thousands of distinct odors, the sense of smell seems to be of little use in observing behaviors, especially those of other humans.

Our kinesthetic sense provides us with the ability to determine differences in weights under various conditions. From this sense, we can observe changes in behaviors under various weight conditions.

For completeness, behavioral observations using the sense of taste are included. However, the sense of taste does not play an important part in behavioral observations.

The unaided senses are most useful in observing the holistic behaviors of individuals. Sensory aids can enhance our senses and allow the observation of behaviors that would otherwise be unobservable. For example, observations of behaviors can be made using microscopes, audio amplifiers, pressure sensors, chemical reaction detectors, heat meters, and the like.

The combined use of the senses is important in making behavioral observations. For example, the behavior of individuals (observed visually) in response to audio communications (observed aurally) is a useful combination of senses for the observation of behavior.

In addition to observing the behaviors of individuals, these methods of observation can also used to observe various phenomena. The three-dimensional capabilities of the visual senses allow observation of distance and depth. Sequential visual observations allow observation and sensing of the passage of time phenomena. The sense of touch provides a means of observing the temperature difference phenomenon, and the vestibular senses provide the means for observing the phenomenon of mass differences.

2.3.2. Behavioral Observables

The methods of observation discussed herein apply to the behaviors of all systems. However, for the purposes of this book, they are intended to develop a quantitative science of living systems with particular emphasis on animals and humans. Therefore, many of the behaviors described herein relate to animals and, specifically, to humans.

Observed behavior can be classified according to basic behavioral types, regardless of the methods of observation. For example, mechanical behavior can be considered as one classification irrespective of the sensor or meter used for its observation.

Mechanical behavior includes mechanical motion of individuals (observed visually), generation of sound waves by individuals (observed aurally), and mechanical pressures such as the pulse (observed by the sense of touch). Mechanical behaviors of individuals are not only the most readily observable but also those most subject to the volitional control of individuals. The mechanical motions of individuals are treated very well by Wells and Luttgens (1976) in their book *Kinesiology: Scientific Basis of Human Motion.* Kelley (1971) also treats mechanical motion in his book *Kinesiology: Fundamental of Motion Description.* Other mechanical behavior is described by Howell and Goldstein (1971) in their book *Engineering Psychology: Current Perspectives in Research.*

Another classification of observables is associated with the chemical behavior of individuals. Chemical behaviors within the body, such as the metabolic process, can be observed with the techniques of chemistry (Aboul Wafa, 1964).

Thermal behavior is another class of observables. This behavior can be observed directly by the sense of touch and indirectly by proper instrumentation such as thermometers and calorimeters, which can measure heat flow, temperature variations, and heat content. The thermal behavior of individuals is almost completely autonomous and is vital to an individual's survival.

A significant class of behavioral phenomena cannot be observed directly through the five senses—electrical behavioral phenomena. Electrocardiograph (EKG), electroencephalograph (EEG), and galvanic skin response (GRS) instruments provide methods for observing electrical behavioral phenomena. The behavior indirectly observed by these instruments includes the automatic electrical pulses that cause contraction of the heart muscle as measured by the electrocardiograph. A treatment of this behavior and its measure is provided by Nelson and Geselowitz (1976) in their book, *The Theoretical Basis of Electrocardiology.* The electrical behaviors of the brain, as determined by the electroencephalograph, are treated by Cooper *et al.* (1980) in *EEG Technology.* The electrical activities associated with muscular activities and changes in skin electrical resistance, as indicated by galvanic skin responses, are treated by Barham and Boersma (1975) in *Orienting Responses in a Selection of Cognitive Task.*

The movement of an electrical charge in an individual is always accompanied by a magnetic field that can be observed so long as sufficiently sensitive magnetic instruments are available. The relatively strong magnetic fields, such as those associated with the heartbeat and the movement of large muscles, have been mapped since the 1960s. Magnetic fields associated with the human brain have been observed (Zimmerman, 1982) using the magnetoencephalogram (MEG).

2.4. Measurement of Behavioral Phenomena

Measurement depends on both a concept of a phenomenon and a method for measuring the phenomenon. As stated, an observer's depth perception allows the concept of distance, one thing being more distant than another. Also, the concept of distance is axiomatic because it is a self-evident truth for any observer. Likewise, the concepts of time, mass, and temperature are self-evident truths for all observers and, therefore, are axiomatic.

As described in Chapter 1, measurement also requires the assignment of some agreed-upon standard unit to the concept where the standard is invariant. These standard units may be either arbitrary or reasoned; generally they are arbitrary. For example, the fundamental unit of distance (length) in the English system of measure was arbitrarily set to be the length of the human foot, whereas in the metric system, the unit of length was intended to be one ten-millionth of the distance measured on a meridian from the equator to the pole and was named the meter. The unit of time is based on the invariant periodicity of the earth's rotation on its axis and its rotation about the sun. The unit of weight in the English system of weights is derived from the weight of a grain of wheat. (there are 7,000 grains in the avoirdupois pound and 5,760 in the apothecaries pound.) There are a number of units for temperature—degree centigrade and degree Fahrenheit being the most common. These two units of temperature are based on the invariant change in volume of specific liquids when heated or cooled.

In addition to the fundamental measures that are based on axiomatic concepts, there are measures derived from these fundamental measures that are important in the measurement and quantification of behavior. These derived measures include mechanical energy, force, pressure, and work—all necessary for measuring and quantifying mechanical behaviors. Derivation of the measures of heat energy is crucial to the measure and quantification of thermal behaviors.

Measurement of nonliving systems behaviors is well known and is the subject of the physical and chemical sciences and their derivative discipline areas such as mechanics and thermodynamics. The focus of this book is the behaviors of living systems, especially those of animals.

Mechanical behaviors of animals with muscles range from large motions, such as running, to very small motions, such as vibration of vocal cords to produce sounds barely detectable by the most sensitive instruments. Wells and Luttgens (1976) document the scientific basis for human motion in their text on kinesiology, which describes the motions of the human body and its parts in terms of physical parameters such as mass, time, length, and velocity. All the laws of mechanics apply to the description and measurement of these observed mechanical behaviors, and all human mechanical behavior can be described, measured, and expressed in terms of energy (Kelley, 1971). For example, Garrett and Reed (1970) developed a computer graphics method for quantifying the kinetic energy of the human body motions; their computer program calculates segmental kinetic energies as well as total body kinetic energy.

The fundamental principle of physics that energy is neither created nor destroyed provides a basis for converting from one energy form to another. Therefore, kinetic energy, which is often expressed in mechanical units, can be expressed in calories when appropriate conversion factors are used. For example, mechanical output energy can be quantified as fractional horsepower over a period of time and can also be quantified in calories. Independent of the energy units used, all mechanical behavior can be precisely described, measured, and quantified by the amount of energy used in the behavior.

Chemical behaviors include metabolic processes, which can be described from the molecular level up to the complete individual. For example, the energy released when a molecule of glucose is metabolized has been precisely quantified by Becker and Hamalainen (1914), and basal metabolism energy rates have been established for human individuals under various conditions by Benedict *et al.* (1919). The chemical reactions of various substances with the human taste buds have been described by Jeppsson (1969), and the energies associated with these reactions can be determined. Likewise, the chemical reactions of aromatics with the olfactory senses have been described by McCartney (1968) and by Denton and Coghlan (1975), and the reaction energies determined. All human chemical behaviors can generally be described in terms of chemical reactions and can be quantified in terms of the energies associated with these reactions.

Thermal behaviors include the heat associated with metabolism and the flow of heat energy to and from the environment. Metabolic energy can be quantified using well-known methodologies described by Consolazio *et al.* (1963). The rate of metabolism is a function of surface area, age, gender, work output, climate, race, and seasonal variations (Becker and Hamalainen, 1914). Voit (1901) determined that the heat output during rest of animals such as the horse, dog, rabbit, mouse, fowl, and man is dependent on their surface area. Rubner (1902) postulated that metabolism is proportional to the superficial area of an animal. For example, the average heat output for the normal human male between the ages of 20 and 40 is 39.5 calories per square meter per hour. All heat processes of the body can be described and measured in terms of energy used. This energy is commonly measured in calories.

Measurements of body electrical behaviors must be performed with specialized and sensitive meters due to the low levels of energy and the small size of the paths through which this energy flows. The electrical activities in the nervous system have been described under various conditions, and electrical waveforms, voltages, and currents of these electrical activities have been measured, allowing the quantification of the energy in these activities (Brazier, 1977). It is theoretically possible to quantify all of the electrical energies in the body.

Behavior in the various classes is measured by different means and involves different types of energy. However, since energy measured in one system of units can be converted to another system of units, it is possible to quantify all the classes in one system of energy units, such as calories. A large amount of an individual's behavioral energy is usually measured in calories. The basal metabolism of an adult weighing 70 kg is approximately 1,750 calories for a 24-hour period and is necessary to maintain body temperature and to support the vital processes such as maintaining the heartbeat, and breathing. Physical work, such as body movement, is the major factor that affects the metabolic rate. For example, Becker and Hamalainen determined that the energy requirements for men in Finland vary from 2,400 calories in 24 hours for tailors to 5,400 calories in 24 hours for men sawing wood.

The above data demonstrates that all objective observations of behavior are based on the utilization of various energy types, that these energies can be directly or indirectly measured, and that the energy measurements can be converted to a common energy measurement system. This leads to the conclusion that behavior can be quantified in terms of the energies used in the behaviors. This conclusion is equally valid for autonomous and volitional behaviors of all animals.

2.5. Available Energy

Inasmuch as energy is neither created nor destroyed, the energies utilized in the behaviors of individuals must be available for these behaviors to occur. One of the most pervasive behaviors of individuals is associated with thermal energy. For example, humans must utilize energy to maintain their body temperature at 98.6°F. To exhibit this behavior, humans must have energy available that can be used to maintain this body temperature. This available energy comes both from the environment in the forms of thermal and solar energy and from food that is used chemically to produce internal heat. If either thermal energy from the environment or food energy falls below that normally required by the individual, the reduced utilization of these energies causes a change in behavior. In the extreme case, a large reduction in thermal energy results in death from hypothermia.

Energy must be available in the form of nutrients for animals to carry out their metabolic processes. Benedict *et al.* (1919) studied the effect of progressively reducing the food intake of normal, healthy human males. Decreased behavior was recorded as a function of reductions in the amount of available energy in the form of food, thus demonstrating the conservation of energy law as it applies to individuals. Thus, an individual's behavior *(b)* is a function of the energy available to the individual (e_a).

The relationship between available energy and those readily observed behaviors of animals (i.e., motion) can be further illustrated by considering the motor unit. A motor unit is composed of a single motoneuron and the muscle fibers attached to the motoneuron. This is a basic behaving element of animals. The readily observed behaviors are the result of one or a number of motor units contracting to cause the observed behavior. A motor unit converts chemical energy into mechanical energy and heat energy each time it contracts to cause an observable behavior. If muscle chemical energy is depleted, as in a fatigued muscle, the muscle does not have energy available to it and therefore cannot contract.

2.6. Summary and Conclusion

1. The behaviors of living systems can be quantified by the energies used in the behaviors.
2. The methods for quantifying the behaviors of living systems are the same as those for nonliving systems (i.e., by the observation of the energies utilized in the behaviors).
3. An observed behavior can occur if, and only if, energy is available to an individual whose behavior is being observed.

CHAPTER 3

Capacity to Direct Energy

3.1. Introduction

We have established that the behaviors of both nonliving and living systems can be measured by the energy in these behaviors. The energies in the behaviors are, in many cases, different from the energies that are available to the systems. This phenomenon can only occur if the available energies are operated on and changed in some way by the system under observation. This phenomenon provides the basis for a concept and a hypothesis that systems have a capacity to operate on (i.e., direct) energies and that this capacity to direct energy is a unique function of a system's structure. That is, the capacity to direct energy is a fundamental behavioral characteristic of a system.

We can test and validate this hypothesis by investigating the relationships between the behaviors of a system and the energies interacting with the system, for both living and nonliving systems. Proof of the hypothesis requires the investigation of systems—from simple molecular systems to the higher animals—to demonstrate the applicability of the concept to both living and nonliving systems.

Simms (1971), in *A Measure of Knowledge*, gives a mathematical treatment of both living and nonliving systems' capacities to direct energy.

The first step in establishing the "capacity to direct energy" as a fundamental behavioral characteristic is to identify the energies in a system's behaviors and the energies in a system's environment. The types of energies that we can observe are described in Chapter 2. The behaviors of systems may include any of these energy types. The environment of a system may also include these energies. Current technology provides a way to measure the amount of energy in a behavior and in a system's environment, and to determine the characteristics of these energies.

The capability to characterize and measure the energies in the behaviors of a system and the energies in the system's environment provides a way to determine how a system operates on the energies in the environment in order to cause a given behavior. The behavior of an animal's motor unit illustrates the way a living system operates on energy to cause behavior. An animal obtains energy from the environment in the form of food. The chemical processes (metabolism) in an animal convert the food energy into chemical energy, which is directly available to the muscle tissue of the motor unit. The structure of the motor unit converts this chemical energy into heat energy and mechanical work. The available chemical energy, the muscle output heat energy, and work can all be measured. The capacity to direct the energy of a particular motor unit can be determined based on its structural characteristics, the available input energy, and the heat and work output energies.

The method illustrated above for determining a motor unit's capacity to direct energy can be used to determine the capacities of animals, plants, and minerals to direct energy. For example, a readily observable behavior of plants is their motion due to growth. There are available energy inputs in the form of food, there are energy outputs in the form of growth motion, and the plant operates on the available energy to produce the growth motion. A plant's capacity to direct energy is a function of its species, which, in turn, is related to the plant's structure. The process for determining a mineral's capacity to direct energy is essentially the same as for animals and plants. For example, the heat energy in the environment of a mineral can be measured, as well as the types and amounts of energy a mineral releases to its environment. Both the input energy and output energy can be measured.

To determine a system's capacity to direct energy we first identify all forms of energy associated with individual systems and their environment. The next step is to identify the directed energies of the individual and its environment that are influenced in some way by the individual. The magnitude of the energies that can be directed by an individual can be determined and quantified. The amount of energy that can be directed by an individual is then the individual's capacity to direct energy.

3.2. General Concepts

General concepts for a substance's capacity to direct energy were synthesized based on energy considerations. A substance is defined as anything that has fundamental or characteristic qualities. A substance can be considered as a particular arrangement of mass and energy that has particular fundamental characteristics. All the mass of a substance is actually energy, as can be seen by the well-known Einstein energy–mass equation ($E = mc^2$). This equation states that energy is equal to mass times the square of the speed of light. A substance can be considered as energy in a particular form and with particular characteristics.

The general concepts were synthesized based on consideration of the energies in a substance and its environment and the exchange of energy between a substance and its environment. An energy system for these considerations was defined as consisting of a substance's energy and the energy in the substance's environment. The total energy in this energy system consists of two separate and mutually exclusive energies, namely the energy of the substance and the energy of the environment. The energy in the environment is all the remainder of the energy in the energy system that does not belong to the substance. To obtain a substance's capacity to direct energy, it is necessary first to compute or measure values for a substance's energy, for environmental energy, and for the energy exchanged between a substance and its environment, and then to determine what energy is directed by the substance.

3.2.1. Substance Energy

The total energy of a substance at any given instant is equal to the sum of all the energies that make up the substance, that is, the energies that the substance possesses. These energies are the energy of the mass that make up the substance and the thermal, chemical, and mechanical energies of the substance.

The energies that make up a substance are internal to the substance. These internal energies can be permanent, semipermanent, or transient. The permanent energies constitute the basic organization of the substance. The semipermanent energies can be stored in a substance and can be used in the process of energy exchange with the environment. The transient energies are those that are rapidly exchanged with the environment.

3.2.2. Environmental Energy

The environment of a substance is defined as the energy surrounding the substance, that is, all the energy that is not in the substance. This includes all the energy that can act on the substance or can be acted upon by the substance. The environmental energy in any energy system includes (1) the energies of all the other substances located in the energy system, and (2) the free energies in the energy system, such as radiant energy. The total energy in the environment at any given instant is the sum of the energies of all the substances that make up the environment plus the sum of all the radiant energy in the environment.

3.2.3. Substance–Environment Energy Exchange

As time changes, the energies in the substance and the environment vary due to the flow of energy between them. The energy accepted by a substance can be in the form of mass, mechanical, thermal, chemical, or radiant energy. Based on the law of energy conservation (i.e., energy is neither created nor destroyed), the energy accepted by a substance must either cause an increase in the substance's energy or it must be released back into the environment. The energy released from the substance into the environment also can be in the form of mass, mechanical, thermal, chemical, and radiant energy.

Substances tend to be in equilibrium with the environment when considered over a reasonably long time. For equilibrium conditions, the amount of energy accepted and the amount of energy released by a substance are equal, and the energy of the substance is not changing. However, for nonequilibrium conditions, a substance can experience changes in its internal energy—for example, during periods of a living system's growth when the accepted energy exceeds the released energy.

The energy accepted by a substance is a function of the type, form, and level of the environmental energy and the substance's characteristics and energy. For example, green plants accept environmental energy in the form of radiant solar energy and carbon dioxide; plants that are not green cannot accept environmental energy in these forms. Animals have the capability of obtaining energy from only certain types of food due to their particular method of metabolizing foods. The process for the release of energy from a substance to the environment is also a function of the energy in the environment and the energy and characteristics of the substance. For example, the release of heat energy from a substance to its environment is a function of the substance's temperature and the temperature of its environment.

3.2.4 Directed Energy

Directed energy is my term for the energy in an energy system that is acted upon by a substance. In general, the energy directed by a substance has been changed from one form of energy to another, or has been controlled in some way. Therefore, directed energy is regulated, controlled, guided, or otherwise processed or changed by a substance.

Directed energies are a subset of energies in both the substance and the environment. Development of formulas for the computation and measurement of directed energy requires a determination of the form of the directed energies in a substance and its environment. Such formulas have been developed by Simms (1971) and are highlighted below.

3.2.4.1. Directed Internal Energy. It might seem that all the energy of a substance at a given time has been operated on in some way and hence is directed energy. However, the energy of a substance includes energies that are not directed by the substance; for example, the mass energy of the nuclear components that make up a substance is not changed or acted upon by the substance. To illustrate this point, consider the following example: A living substance may be composed of many atoms of various types of matter. During the lifetime of the living substance, the mass energy of the atoms making up the living substance is not changed or altered in any way, that is, the atomic energy of the substance ($E = mc^2$) has not been changed or directed by the substance.

The nature of the internal energy of a substance can be seen by considering one of the simplest molecules, a diatomic molecule. Physics (Richtmyer and Kennard, 1947) and chemistry (Prutton and Marion, 1947) principles state that the total energy of a diatomic molecule consists of three parts: (1) the energy of the original atoms, (2) coulombic energy resulting from the electrostatic attraction between the electrons and the nuclei, and (3) exchange energy. As mentioned above, the mass energy of the original atoms is not directed energy. The solution to the appropriate wave mechanics equations gives the total energy of a diatomic molecule as a function of separation distance between the two nuclei. As the nuclei are brought together, the attraction between the nuclei increases until a point is reached where the attraction is maximum. A stable molecule will have a specific separation distance between the two nuclei, the distance being different for different diatomic molecules. The separation distance between the two nuclei is associated with a particular electronic (coulombic) energy. When the molecule is in its lowest energy state of zero degrees absolute temperature, there is some remaining electronic energy in the molecule. This remaining electronic energy is associated with the formation of the molecule but is not directed by the molecule, just as atomic mass energy is not directed by the molecule.

The total energy of a molecule also includes vibrational and rotational energy and the energy of thermal motion. The lowest energy state of a molecule occurs when there is no energy in the molecule's environment. Any energy in the environment that is accepted by a molecule will cause an increase in its internal energy. These increased internal energies have been operated on by the substance and are therefore directed energies.

The internal energies of polyatomic molecules are more complicated than those of diatomic molecules. However, the principles of internal energy and directed internal energy apply equally well to diatomic and polyatomic molecules. More complex substances such as aggregates of molecules, living cells, tissue, organs, and organisms exhibit much more complicated internal energy arrangements. The directed internal energy of complex substances is discussed in more detail in the following sections.

3.2.4.2. Directed Environmental Energy. Many substances are capable of directing some of the forms of environmental energy previously described. The amount and form of this directed environmental energy are determined by the type of substance being considered. Because directed energy is dependent on the type and form of the substance, general directed environmental energy relations applicable to all substances are precluded.

Substances direct those energies in the environment that impinge upon the substance but are rejected by the substance. This rejected energy is directed by the substance because the substance causes a change in direction of energy flow. An example of this directed energy is solar radiant energy which is reflected by a substance. The amount of directed environmental energy can be quantified when the substance and the local environment are specified.

Some substances can direct energies such as heat and wind. These directed energies can be treated in the same manner as rejected radiant energy.

3.2.4.3. Directed Exchange Energy. Substances are continually accepting energies from their environment and releasing some of their energies into the environment. The characteristics of a substance determine which environmental energies are accepted into the substance. As noted previously, a basic characteristic of green plants is their acceptance of solar energy. The solar energy accepted by these plants has been operated on by these green plants and thereby is directed exchange energy. In general, any environmental energy accepted by a substance has been operated on by the substance and is directed exchange energy.

Substances are continually releasing energy into the environment. For example, the byproducts of plant and animal metabolism are continually being released to the environment by them. These byproducts have been operated on and, therefore, directed by plants and animals. A more universal example is the release of heat energy from minerals, plants, and animals into a cool environment. This released heat energy has been operated on and is thereby directed exchange energy.

3.2.4.4. Total Directed Energy. The total directed energy of a substance is the combination of the directed energies in the substance and in the environment and in the directed energies exchanged between a substance and its environment. Since directed energy is a function of the specific characteristics of a substance, a detailed development of the directed energy of a substance must wait until the various kinds of substances are considered. However, some general characteristics of substances can be identified based on energy considerations.

3.2.5. General Characteristics of Substances

Because a substance's capacity to direct energy is a function of its basic characteristics, these characteristics need to be better understood and are considered in this section.

The most basic characteristics of any substance are the minimum energy and capacity to direct energy required for the substance to maintain its identity, that is, to survive. The capacity to direct energy and the energy required for survival vary from substance to substance. The minimum capacity to direct energy and the energy required for a substance to survive are, in general, well defined and are characteristics of the substance. Substances can be considered by their organizational structures and the energy required for survival.

The basic substances of the universe, such as the physical elements, can survive in the absence of energy from the environment. The directed energy required for the survival of these elements is the energy associated with their minimum electronic energy. This ground state energy is the lowest energy level at which any substance can exist.

Complex substances such as plants and animals require rather elaborate energy systems and methods of exchanging energy with the environment. These substances must not only have ground state energy and a thermal energy exchange with the environment, but also a mass exchange with the environment in order to survive. These complex processes for controlling energy allow these substances to obtain sufficient energy from the environment for survival; however, after a period of time, plants and animals die from "natural causes." The minimum energy required for these substances to survive is rather well defined.

To survive, substances must have a sufficient capacity to reject energies which exceed values that would cause the substance to change state or cease to exist. The energy level that causes a substance to expire is a basic characteristic of the particular substance. Energy in different forms will have different effects on a substance depending on such basic properties of the substance as selective rejection of energies at its surface, selective transmission of energies through the substance, and selective absorption of energies by the substance. The maximum levels of the various types of energy—radiant, mechanical, electrical, chemical— that cause a substance to expire can be determined. If a substance is to survive in a given environment, it must be capable of rejecting all energy above the maximum tolerable level.

To survive, all substances must possess a sufficient capacity to (1) direct the minimum internal energy characteristic of the particular substance, (2) direct energy exchange with the environment to maintain the substance's internal energy, and (3) reject energy at the interface with the environment that will exceed the maximum energy capability of the substance.

The conditions of the energy exchange at the interface are such that, in general, there must be an energy balance between accepted and rejected energies. Otherwise, survival energy conditions are not maintained. The survival capacity to direct energy of a substance is determined by obtaining all the directed energy required for the substance to survive—this is the substance's minimum capacity to direct energy.

When energy takes form (i.e., is organized into a substance), it acquires a fundamental characteristic, which is a capacity to direct energy.

3.3. Minerals

A brief review of the important characteristics of minerals provides background for the development of a mineral's capacity to direct energy. The review is limited to those characteristics which assist in the identification of a particular mineral and those characteristics which allow the determination of a mineral's energy and its capacity to direct that energy. The standard classifications and characteristics of mineralogy, chemistry, and physics are used to identify these characteristics. The capacity to direct energy, as it applies to minerals, is a characteristic associated with the behaviors of minerals with respect to their own energies and those of their environment.

3.3.1. Characteristics of Minerals

Essential mineral characteristics are those relating to chemical composition, crystalline form, crystallo-physical properties, and specific gravity. These are identical, or vary only within certain defined limits, in all specimens of the same mineral. The characteristics essential to a mineral's capacity to direct energy are those that can be expressed numerically.

3.3.1.1. Optical Characteristics. The action of crystallized matter on transmitted light is a very important characteristic in mineralogy. The refractive indices, optic axial angle, strength of double refraction, and extinction angles on certain faces are constant for each mineral. These characteristics are important from the point of view of the acceptance or rejection of light energies.

3.3.1.2. Thermal Characteristics. The specific heat and melting point of minerals are essential characteristics that can be measured exactly and expressed numerically and, therefore, are of primary importance in the determination of a mineral's capacity to direct energy.

3.3.1.3. Characteristics Depending on Cohesion. Hardness, or the resistance that a mineral offers to being scratched by a harder body, is an indicator of the ability to accept or reject environmental energy. In addition, parting planes, etching figures, and pressure and percussion figures are sometimes important in describing and distinguishing minerals.

3.3.1.4. Specific Gravity. The density, or specific gravity, of minerals is an essential characteristic of considerable deterministic value. In minerals of constant composition it has a definite value and can be expressed numerically.

3.3.1.5. Chemical Characteristics. Chemical composition is an important characteristic of minerals. It is the basis of all modern systems of classification. However, a mineral cannot be defined by chemical composition alone since many minerals of the same chemical composition are dimorphous or polymorphous. In such cases, other essential characteristics are required to define a mineral.

3.3.1.6. Nomenclature and Classification of Minerals. A mineral is completely defined by the specification of its chemical composition and crystalline form. When dealing with a definite chemical compound, defining minerals is relatively easy; thus, corundum, cassiterite, and galena are well-defined minerals. However, with isomorphous mixtures, the division into species and into subspecies and varieties must be, to a certain extent, arbitrary as there are no sharp lines of demarcation in many isomorphous groups of minerals.

Minerals may be classified in various ways to suit different purposes, for example, according to their uses, modes of occurrence, system of crystallization, etc. For purposes of determining the capacity to direct energy characteristics, two major methods of classifying minerals are important: (1) by the essential characteristics of a substance which identify it as being different from other substances and by those characteristics which allow identification of major changes in state, and (2) by the energy of a mineral and the characteristics which determine its energy and the energy interface with its environment.

The most important essential characteristics are: (1) mass (in terms of some standard such as specific gravity or density), (2) thermal capacity, (3) dissociation energy level, (4) sublimation point, (5) structural organization (the atomic, molecular, crystalline, and aggregate organization of the substance), and (6) the surface absorption and reflection characteristics (the capability of exchanging energy with the environment).

3.3.2. Internal Energy

Because this book focuses on the capacity of individuals to direct energy, a mineral individual must be defined. To be consistent with a definition of individuals in the plant and animal kingdoms, a mineral individual must be the smallest recognizable unit which retains its essential characteristics. A molecule is the smallest unit of matter which can exist by itself and retain all the properties of the original substance. Therefore, a mineral individual is a molecule of a particular mineral.

The previously identified internal energies of a substance can be tailored for application to a mineral individual (a molecule). These energies are: (1) the mass energy of the molecule, (2) the electronic energy of the molecule, (3) the translational energy of the molecule, (4) the molecular vibrational energy, and (5) the molecular rotational energy.

The energy due to the mass of the molecule is determined by the formula $E = mc^2$, where m is now the mass of the molecule. The electronic energy results from the location and motion of the electrons about the nuclei of the atoms which constitute the molecule. The translational energy is due to the kinetic energy of the molecule with respect to some reference system. The vibrational energy is caused by the vibration of the atoms of a molecule with respect to one another. The rotational energy derives from the rotation of the atoms of the molecule about the rotational axis of the molecule. All these energies can be quantified using the methods and procedures of the physical sciences.

3.3.3. Environmental Energy

A mineral individual's environment includes individuals of its own kind, individuals of other substances, and radiant energy. The energies of these environmental components combine to form an individual's energy environment. The energy of a specific environment can be determined using equations developed by Simms (1971).

The radiant energy in the environment can be specified in terms of intensity, spectral distribution, or direction of travel. The total radiant energy in the environment is a combination of the various radiant energies in the environment.

3.3.4. Mineral–Environment Energy Exchange

Mineral individuals are continually exchanging energy with their environment. To determine the exchange energy that is directed, it is first necessary to determine the forms, types, and amounts of energy being exchanged between a mineral and its environment.

A mineral molecule can obtain energy from its environment by the absorption of radiant energy and by collision with atomic particles, which causes an interchange of energy. A molecule can transmit energy to its environment by radiant energy and by collisions. The abilities to absorb and emit energy under various conditions are essential characteristics of minerals. For instance, the molecules of a given mineral absorb and emit radiant energy only at specific wavelengths and then only under certain energy conditions. The amount of energy a molecule absorbs depends on its internal energy state. The amount of energy a molecule releases is a function of its internal energy state and the amount of energy being absorbed.

To quantify the energy flow through a mineral individual, it is necessary to determine the absorbed and re-emitted energy as a function of the internal energy conditions of the individual.

All mineral molecules tend to emit energy in an attempt to reach their ground state, which is their lowest energy state. On earth, the lowest energy state is not attained because energy from the earth environment is accepted by the molecules. This environmental energy maintains the molecules in energy states that are higher than their lowest energy states. The rate of emission is a function of many things, including the internal energy level of the mineral, thermal properties of the mineral, surface conditions of the mineral, thermal coupling between the mineral and the environment, type of material in the local environment, energy state of the environmental materials, and thermal properties of the materials in the environment. Emitted energy results in the individual's internal energy being reduced by its changing to a lower quantum energy state. If the individual is in an environment of other individuals which are rather widely separated (such as a gas), then the energy can be emitted by both radiation and collision, the energy released by collision being transferred to the other individuals. If the individual is in an environment of similar individuals that are in close proximity and hence mutually affect one another—such as in a solid material—then the energy is transferred among individuals by other mechanisms.

The exchange energies just described can be quantified using the methods and procedures of physics. The equations allowing the quantification of these exchange energies are documented elsewhere (Simms, 1971).

3.3.5. Directed Energy

A mineral individual's directed energy is obtained by determining which of the mineral's internal, environmental, and exchange energies described above are directed by the individual. The total energy directed by the individual is determined by combining these energies.

3.3.5.1. Directed Internal Energy. Directed internal energy is a subset of all the internal energy of a mineral individual. Therefore, the question is: What portion of an individual's internal energies is directed? It has already been determined that the internal mass energy of any substance is not directed; therefore, the mass energy of a mineral individual is not available and cannot be directed by the individual.

An investigation was made to determine if the electronic energy of a mineral molecule in its ground state could be directed by the individual. The energy of a mineral molecule is minimal when in its ground state. It was determined that the ground state electronic energy is not directed by an individual. A detailed treatment of a mineral's ground state electronic energy is reported in *A Measure of Knowledge* (Simms, 1971).

The translational energy of a molecule as well as the molecular vibrational and rotational energies are internal energy states of a molecule. These states change in response to the heat energies in the environment. From another viewpoint, these internal energy states are different internal structural arrangements or organizations of a molecule that accommodate different internal energy levels. A molecule changes its internal structural configuration and thereby processes or changes energy; by my definition of directed energy, this internal energy is directed.

3.3.5.2. Directed Environmental Energy. For a mineral individual to direct environmental energy, it must act upon, convert, or in some way change the energy in the environment. This occurs when environmental radiant energy impinges on a mineral and is caused to change its direction at the surface of the mineral (e.g., radiant energy which is reflected by the mineral). Also, the change in direction of a particle due to a collision with an individual is another form of directed environmental energy. The environmental energy that impinges upon an individual and is reflected back to the environment is directed energy.

The ability to reject energy is an essential characteristic of a mineral. According to quantum mechanics theory, a mineral rejects thermal energy unless this energy is directly related to a number known as Planck's constant. Rejected energy is also a function of the particular physical characteristics of molecules.

The directed environmental energy for a mineral individual was found to be the energy rejected at the surface of the individual. This rejected energy is composed of radiant energy and redirected kinetic energy.

3.3.5.3. Directed Exchange Energy. The mineral–environmental energy exchange described above can be summarized as follows: (1) energy is accepted from the environment and increases the energy state of the individual, and (2) energy is emitted to the environment by an individual, thereby decreasing its energy state. This exchange energy has been operated on by the mineral and therefore is directed energy. Directed exchange energy is a function of the particular characteristics of a mineral individual. The absorption and emission of radiant energy are essential characteristics of a particular mineral species, and the spectral nature of the absorbed and emitted radiant energy is unique to this particular mineral species.

The amount of energy directed can be determined by considering the condition where the mineral is in energy balance (equilibrium) with the environment. Under these conditions, the energy accepted by the mineral substance is equal to the energy emitted by the substance. For equilibrium conditions, there is a steady flow of energy through the substance and there is a particular internal energy condition associated with this flow of energy. Since all the energy flowing through the substance is operated upon by the substance, it is, by definition, directed energy. The amount of this energy flowing through the substance can be obtained from equations developed in *A Measure of Knowledge*.

3.3.5.4. Total Capacity to Direct Energy. The total directed energy is obtained by combining the directed internal, exchange, and environmental energies. These directed energies can be quantified using the techniques described in the preceding sections and the equations developed in *A Measure of Knowledge*.

As described above, it is a mineral's structure and organization that gives it a capacity to direct energy. These structural and organizational characteristics of minerals give them the capacity to direct specific types and amounts of energy.

For example, a given mineral molecule has a capacity to direct thermal energy ranging from zero up to the amount of energy that will cause disassociation of the molecule. Disassociation of a molecule into its atomic constituents signifies the death of the molecule since it no longer exhibits the characteristics of the mineral molecule.

Although a mineral molecule has a capacity to direct various energies, its environment may contain only a few types of energy. Under these circumstances, the behaviors of the mineral molecule are related only to the part of its total capacity to direct energy that operates on the environment.

The conceptual and theoretical development of a mineral's capacity to direct energy described above is derived from, or supported by, a wealth of experimental data. A current *Handbook of Physics and Chemistry* provides data on all aspects of the behaviors of minerals under various environmental energy conditions. These mineral data include, for example, the spectral lines at which various minerals emit energy, the types and amounts of energy that are absorbed by a mineral, the types and amounts of energy that are reflected from minerals, and the internal energy levels of minerals when subjected to various energy environments.

We have "proven," then, in the preceding sections that minerals have a capacity to direct energy and that tools and data exist for quantifying these capacities to direct energy.

3.4. Plants

A major characteristic differentiating plants from minerals is life. To extend the concept of a substance's capacity to direct energy to plants, it was essential to have an adequate definition of life. The necessary and sufficient conditions for a substance to be recognized as living are that it be a discrete mass of matter with a definite structure or boundary, undergoing continual interchange of material with its environment without manifest alteration of basic characteristics over short periods of time, and originating by some process of division or fractionation from one or two preexisting organisms of the same kind.

The continual exchange of material with the environment is the metabolic criterion for life, and the origin from preexisting systems is the reproductive criterion for life.

The above definition of life is satisfactory for the author's purposes. Although philosophers have, from the earliest time, attempted to define life, not one completely satisfactory definition has ever been advanced. The difficulty of their task can be appreciated because it becomes increasingly evident that there is no clear line of demarcation between the living and the nonliving. Nevertheless, as Claude Bernard* indicated in his essay called the "Phenomena of Life," there are a number of properties of living matter which, taken collectively, serve as a rough and ready means of differentiating life from relatively inert systems. These are:

- Assimilation and respiration
- Reproduction
- Growth and development
- Movement
- Secretion and excretion

Considered singly, not one of these is, in itself, characteristic of living matter alone. However, as yet, no system that could be considered as nonliving has been found to exhibit all of these properties together.

Today, life is best characterized and distinguished from nonliving systems by the processes that living organisms carry out. Two key processes characterize all living systems. One of these is metabolism, the sum of all the chemical reactions taking place in an organism. The second is reproduction, the replication of themselves (Purves *et al.*, 1992).

3.4.1. Characteristics of Plants

The changing of food into living matter is called assimilation. All plants have the capability to assimilate material and convert the material into energy forms required for the plant to live. Green plants have the capability of synthesizing complex organic substances from carbon dioxide, water, and simple inorganic salts such as sulfides, nitrates, and phosphates. Non-green plants break down food into forms of energy the plant can use.

The term *respiration* is used in biology and biochemistry to describe the oxidation of material to support the life processes. Respiration is a fundamental characteristic of plants.

The sum total of the activities, maintenance, repair, and growth in living organisms is known as metabolism. The mechanisms and process of metabolism are essential characteristics of plants, and certain parts of the process may be unique to a particular species. The metabolic machinery by which chemical changes are brought about consists of enzymes, organic catalysts for a specific chemical reaction.

The ability to reproduce is an essential characteristic of all plants. Everything that lives will die sooner or later, as opposed to nonliving systems which may never die or change form. However, a plant species continues to live through the reproduction process.

* Cited in *Biochemistry, Encyclopaedia Britannica*, Vol. 3, 1951. Bernard's views and statements are found in the 17 volumes of his published lectures.

Growth, as used herein, takes place only through the activity of protoplasm. As such, growth is essentially an irreversible process and involves more than just an increase in size or weight. This definition of growth precludes increases in weight due to absorption of water being considered as growth.

Growth and development are essential characteristics of plants. All plants start from one cell during each generation and must grow in order to reach maturity. The mass of a plant varies as a function of the growth process. This is in contrast to minerals where the mass of a substance is constant and can be used as an indicator of its energy.

Living matter has been named protoplasm by biologists. Most biologists believe that protoplasm owes its living nature to its organization—rather than to the possession of some particular vital element—and can be explained in physical and chemical terms. The mass of a plant is a good indication of its energy. The method of growth is an important characteristic of plants.

Plants have movement due to growth and in some cases in response to stimuli from the environment, for example, a Venus flytrap responding to an insect landing on its petals.

To determine a given plant's capacity to direct energy, it is necessary to distinguish between the various kinds of plants. The classification of plants is based mainly on reproductive structures and behavior with the vegetative characteristics being of secondary importance. This method has gained favor because the reproductive characteristics are less likely than the vegetable characteristics to be influenced by environmental factors.

The species is the basic unit of organism classification. A species is a kind of living organism (e.g., a white oak, a sugar maple, etc.). A species may be defined technically as usually the smallest unit in the classification system. It is a group of individuals of the same ancestry, of similar structure and behavior, and of stability in nature; that is, the members of a species retain their characteristic features through many generations under natural conditions.

The essential characteristics for the determination of a plant's capacity to direct energy relate to (1) mass of the plant, (2) the temperature of the plant, (3) metabolism, (4) method of energy transfer between the plant and the environment, (5) methods of extracting energy from the environment, (6) work done on the environment, and (7) structural organization of a plant.

3.4.2. Internal Energy

In plants, as in minerals, we are interested in determining the capacity to direct energy for the smallest unit which still possesses the essential characteristics. For plants, the smallest unit is an individual organism of a given species. The sizes of plant individuals vary from microscopic one-cell plant species to large organisms such as the redwood and giant sequoia trees.

To ascertain the capacity to direct energy of an individual plant, it is necessary to determine the energy of the individual and the subset of this energy that is directed by the individual. A plant's energy can be considered in terms of the constituent elements of the plant. The characteristics and types of the elements which constitute a plant are either known or can be determined. Therefore, the energy of a plant can be calculated by determining the energy level of each element and summing these energies.

The internal energies of a plant include mass, ground state electronic, and heat energies. However, this does not account for all the energies of a plant. The additional energies of the plant are due to the mechanical motion of certain molecules of the plant with respect to the general coordinates of the plant, and to the energy of protoplasm formation. The mechanical motion of certain molecules with respect to most of the plant molecules is due to such factors as capillary action, different concentrations of solutions caused by metabolic process, transpiration, and the like.

If a coordinate system is selected which is centered on the plant, then the majority of the molecules which constitute the plant will be stationary with respect to this coordinate system. There will be certain short-term movements due to factors such as thermal agitation. But on a long-term basis, the molecules can be considered to be stationary with respect to a plant-centered coordinate system. However, other molecules will have motion with respect to this coordinate system. For instance, the metabolic process causes different concentrations of materials within the plant which result in a flow of metabolic byproducts out of the plant. An example of this motion is the flow of sap in those plants with veins.

When a plant assimilates food into protoplasm, it stores chemical energy. In addition, when green plants synthesize complex organic substances from inorganic material and solar energy, they increase their chemical energy. In these processes, energy is supplied from some external source, such as solar energy, to cause the formation of the complex organic materials. Because these resulting complex materials are internal to the plant, their chemical energies of formation are considered to be stored as internal energy of the plant. Indeed, if these materials were to be dissociated through oxidation, the formation energies would be released. During the growth of a plant, protoplasm is formed into structure. This formation process continues as long as the plant grows. The amount of formation energy is an indicator of a plant's structure and organization. For example, the formation energy and organization of a single-cell plant are smaller than the formation energy and organization of a large multicellular plant.

At any time in its life, a plant is composed of a particular mass of various living materials. If these materials were burned (by the catabolism process), a specific amount of energy would be released. The energy released in the burning process is the energy of dissociation of the plant and also represents the energy required to form the plant from raw material, that is, a plant's organizational energy.

The amount of energy released by the metabolic burning of materials, such as carbohydrates, proteins, and certain fats, has been experimentally determined and documented in the robust literature on nutrition. With these values and the composition of a plant, the plant's organizational energy can be determined.

Plants have the capability of internally storing foodstuff in the form of sugars, carbohydrates, and fats. The energies in these stored foodstuffs can be converted by a plant into heat and mechanical energy. The amount of this stored energy is a function of the amount of foodstuff, the type of foodstuff stored, and the efficiency of the plant in converting a specific type of foodstuff into a usable form of energy.

The metabolism of internally stored foodstuff is carried out in approximately the same way in all plants. The chemical equation for the reduction of a given foodstuff is similar in all plants and the heat of reaction is the same. By knowing the chemical process and the amount and type of stored foodstuff, the amount of energy released through metabolism can be determined. The rate of metabolizing stored energy is governed by factors such as temperature, rate, and method of disposing of metabolism byproducts.

The total energy in a plant is the combination of the following:

• Mass energy
• Ground state electronic energy
• Chemical energy
• Heat energies
• Miscellaneous energies due to metabolism, capillary actions, etc.
• Energies that can be released from stored foodstuff

3.4.3. Environmental Energy

The local environment of a plant is more restrictive than that for minerals because a plant cannot exist in as wide a range of energy conditions as a mineral. That is, the environment that can be tolerated by plants is restricted—extreme temperatures kill plants.

The characteristics of plant species cause them to adapt very slowly to changes in their local environment. Therefore, they cannot, in general, exist in a local environment that changes drastically in a short period of time. Except for the requirement for a somewhat restrictive environment, the local environmental energy of a plant is the same as that considered in the general environment discussed above and that considered for minerals.

3.4.4. Exchange Energy

The internal energy of a plant can be maintained only by an exchange of energy with its environment. This continual exchange of energy permits the plant to be in energy equilibrium with its environment. To determine the amount of the exchange energy that is directed, the total amount of exchange energy first must be determined. The total energy exchanged between a plant and its environment is the energy accepted by the plant and the energy released by the plant into its environment.

Because there are significant differences between the energy accepted by green and non-green plants, their exchange energies cannot be treated the same. The major difference is that green plants accept solar energy directly whereas non-green plants cannot. The acceptance of energy by green plants is treated first.

Energies accepted by green plants include those associated with photosynthesis. This process is fundamental to all life because it describes the synthesis of food for all living systems. The summary chemical equation for the photosynthesis process for all green plants is:

$$6CO_2 + 6H_2O + \text{energy} \rightarrow C_6H_{12}O_6 + 6O_2 \qquad (3.1)$$

By this equation, the accepted mass is six molecules of carbon dioxide (CO_2) and six molecules of water (H_2), and the accepted energy is radiant solar energy from the sun. A green plant converts this accepted mass and energy into glucose and oxygen. The chemical energies and solar energy in this equation are known and can be quantified.

Non-green plants do not synthesize their food the way green plants do. They must accept their food energy from the environment in the form of organic biochemicals. Usually their food is in some form of glucose that has been synthesized previously by green plants.

Both green and non-green plants accept chemicals necessary for metabolism as well as energy in the form of heat (thermal energy) from their environment.

All plants release the byproducts of metabolism into their environments. These byproducts have mass and chemical energies. The basic chemical equation for plant respiration metabolism is:

$$C_6H_{12}O_6 + 6O_2 \rightarrow 6CO_2 + 6H_2O + \text{energy} \qquad (3.2)$$

The byproducts of this chemical reaction are six molecules of carbon dioxide (CO_2) and six molecules of water and chemical energy. Both carbon dioxide and water are transferred to the environment. The chemical energy is used internally by the plant to drive biochemical reactions. The chemical energies associated with the byproducts of the respiration equation are well known and have been quantified.

The amount of energy a specific plant exchanges with its environment is a function of (1) the plant species, based on its size, energy acceptance and rejection characteristics, and where it can live; (2) the actual size and age of a plant; (3) the surface area of the plant; (4) the area of a plant's photosynthesis structure; (5) the energy amount and form in a plant; and (6) the energy amount and form in the environment.

The rate at which the metabolic process releases materials into the environment is a function of the temperature of the plant and the environment. In general, the metabolic process of plants stops below zero degrees centigrade, increases as the temperature increases above zero to some maximum value, which is usually at approximately the average temperature of the local environment. The metabolic process then decreases with increasing temperature until a temperature is reached which kills the plant by stopping the metabolic process. The metabolic rate is also a function of the plant species. Although the basic chemical reaction for the metabolic process is similar in all plants, individual differences, such as cell permeability to various solutions, can occur among plant species.

The rate at which a mature plant releases energy to the environment due to metabolism can be determined from the above equation and the metabolic rate. The rate of mass transferred to the environment is equal to the mass of carbon dioxide and the water resulting from the catabolism of one $C_6H_{12}O_6$ (glucose) molecule times the catabolism rate in the molecules of glucose per second.

The energy transferred to the environment from a plant is obtained by combining the energies due to radiation, conduction, and catabolism. If it is assumed that the catabolism process does not materially affect the temperature of the plant, the energy of catabolism can be considered independently of the radiation and conduction energy, thus allowing it to be added to the radiation and conduction energies. The energy flow to the environment varies with time because of the changing internal and environmental conditions and the particular stage of the plant's development. However, for a mature plant, the average energy output rate must equal the plant's average rate of accepting energy. Therefore, in general, the plant can be considered to be in energy equilibrium with the environment so that the flow of energy to the environment is balanced by an equal flow rate from the environment to the plant (i.e., accepted energy). Under these equilibrium conditions, the exchange energy can be determined. Short-term variations and non-equilibrium conditions can be determined by considering the differences between the emitted and accepted energy of a plant and the characteristics of the plant.

3.4.5. Directed Internal Energy

To obtain the internal capacity to direct energy of a plant, the portion of the internal energy that is directed by the plant must be determined. The various internal plant energies are considered in the following sections to determine those particular energies that are directed.

The atomic energy due to the mass of the plant is not directed in any way. This energy, as represented by the term mc^2, is not changed or utilized by the plant in any manner.

It was previously demonstrated that the ground state potential energy of a mineral molecule was not directed energy. Since a plant is composed of molecules made from elements and mineral molecules, it follows that the ground state energy of the molecules making up a plant is not directed by the plant. Independent of this argument, plants die at temperatures well above the temperatures associated with ground state energies. Therefore, the ground state energy is not directed plant energy.

The internal heat energy of a plant is a function of the plant's mass and its specific heat. If the mass and the specific heat of a plant can be determined at some time, then the internal heat energy of the plant can be quantified. Some of the internal heat energy must be directed by the plant to maintain a thermal relationship with the environment. The amount of heat energy directed by a plant depends on the temperature range in which the plant can exist. For example, if a plant cannot exist below a given temperature, then the remaining heat between this point and absolute zero cannot be considered as being directed by the plant. By the same reasoning, if a plant cannot exist above a given temperature, then the heat energy between this temperature and the temperature at which the molecules of the plant dissociate cannot be considered as being directed by the plant. The temperature range over which a plant can exist may vary widely among plant species. Some plants have been known to exist at temperatures in the neighborhood of absolute zero. Existence is defined here as the capability to meet the metabolic and reproductive criteria in a normal temperature environment after having been subjected to the low temperatures.

The energy that can be obtained by a plant from internally stored foodstuffs is all directed energy. Food can be stored in forms such as sugars, starches, and fats. A plant has the capability of extracting a specific amount of energy from food, depending upon the type of food. The energy obtained from food by the catabolic process has been processed or changed by the plant and therefore is directed energy.

The mechanical energy of a plant is a result of its metabolic process and its structural organization; therefore, this mechanical energy is directed. To illustrate, the energy resulting from the movement of liquids through a plant due to osmosis, various differences in chemical composition, and capillary action are all due to the functions and organization of the plant. Further, these energies would not be present if the plant did not exist. The amount of energy is a function of the particular organizational structure of a given plant species. The form and type of this energy are functions of the type of plant species. For example, the capillary action in large plants with veins differs greatly from this action in small, one-cell plants without veins.

3.4.6. Directed Environmental Energy

Plants can direct environmental energy only in a restricted sense. They can reject radiant energy and other energy impinging on them from the environment, and, through the process of growth, a plant can displace mass in the environment. The energies rejected by plants can be identified and measured. The methods that plants utilize in rejecting energy are a function of a particular plant species.

A plant directs environmental energy when it displaces soil in the growth process. The amount of energy directed by this process is a function of the pressure that can be generated in the plant by the growth process. The amount of pressure and the amount of displacement due to growth can be determined at any given time. Therefore, the amount of directed environmental energy can be determined.

3.4.7. Directed Exchange Energy

The exchange energy between a plant and its environment has been operated on in some manner by the plant. Therefore, all this exchange energy has been directed by the plant.

3.4.8. Total Capacity to Direct Energy

The capacity of a specific plant individual to direct energy can be obtained by summing the directed internal, exchange, and environmental energy to yield the total directed energy.

3.5. Animals

The energies available to animals can be vastly different from their behavioral energies, as evidenced by the available chemical energy in the form of food which an animal uses to produce mechanical and heat behavioral energies. Individuals are capable of absorbing radiant energy in the thermal and solar spectrums and converting the energy into body heat. The conversion is from one energy form to another and results from the individual directing (i.e., operating upon, changing, or controlling) the impinging energies. Individuals also reflect electromagnetic energy and thereby direct this energy. Chemical energy in the form of food is, of course, one of the most important energies available for animals. An individual directs many types of energy with respect to food. These directed energies include mechanical energy to obtain and ingest food, chemical energy to perform digestive and metabolic processes, heat energies to maintain body temperature, and chemical and mechanical energies to perform the muscular activities of the body. The amount of energy that can be directed by an individual—its capacity to direct energy—is a function of its genetic endowment and its learned behavior. The genetic differences between pygmies and Caucasians, for example, provide them with different radiated thermal energy capabilities (radiated thermal energy is a function of total area of the individual) and different metabolic rates.

An individual's capacity to direct energy can change for many reasons. Growing to adulthood changes the individual's total body area and thereby its capacity to direct thermal energy. Growth also results in better muscular development and therefore increases the individual's capacity to direct mechanical energy. Debilitating diseases, such as metabolic disorders and muscular dystrophy, cause a reduction in an individual's capacity to direct energy. As an individual's capacity to direct energy increases, the behavioral energy changes in the same way, given adequate available energy. Because the capacity to direct energy decreases due to things such as a debilitating disease, behavioral energy also decreases.

The readily observable behaviors of animals (i.e., the motions of animals) are the result of contraction of muscle motor units. There are many types of motor units, depending on the types of muscle fibers in the motor units and the number of muscle fibers. The different structures of motor units give them their own unique capacities to direct energy. As an example, the motor units which cause locomotion of an animal are very different from those that make up the "smooth" muscles that cause food to flow through the digestive tract of animals. Each motor unit converts available energy, in the form of chemical energy, into the mechanical motion that is observed as the animal's behavior.

In *A Measure of Knowledge*, mathematical equations are developed for an animal's capacity to direct energies in terms of the types of energies that are directed. The following is a summation of the concept of an animal's capacity to direct energy as developed in *A Measure of Knowledge*.

3.5.1. Characteristics of Animals

The characteristics of animals are important to the concept of an animal's capacity to direct energy because they assist in differentiating one species from another. It is also important to know physiological and morphological characteristics in order to assess the degree of an animal's organization.

3.5.1.1. Physiological Characteristics. Animals cannot synthesize their food but instead must depend on the preexistence of organic substances. Because these organic substances are the product of other living substances, the animal is essentially a parasite on existing plant and animal life. Another general characteristic of animals is the growth and size limitation. A given member of an animal species obtains a particular size at maturity and then maintains approximately this size for the remainder of its life.

We are interested in the physiological characteristics which relate to the transfer of energy between an animal and its environment, the energy transformations internal to the animal, and those energy transformations in the environment which are caused by the animal.

3.5.1.2. Physical. All animals must obtain physical substances from the environment. These substances are both inorganic and organic. The organic material is the animal's food; the inorganic materials include oxygen, water, and certain minerals. In addition, all animals must have a thermal energy exchange with the environment. Cold-blooded animals obtain thermal energy from the environment for survival. Warm-blooded animals are not quite so dependent on the environment's thermal energy.

The methods for obtaining physical substances from their environment vary depending on the animal species. The final usage of the physical substances is in the individual cells where the processes are essentially the same for all animals.

3.5.1.3. Chemical. The chemical process involved in changing materials from the environment into other forms and into energy is normally referred to as metabolism. Temperature, moisture, atmosphere, nutrients, and many other environmental factors affect the process of metabolism. The primary metabolic process is essentially the same in all animals. That is, organic foods are reduced by combining with oxygen (burning). Indeed, the fundamental process of respiration is the same for all living things, plant or animal. The fundamental process for the metabolism of carbohydrates is given by the summary chemical equation:

$$C_6H_{12}O_2 + 6O_2 \rightarrow 6CO_2 + 6H_2O + \text{energy} \tag{3.3}$$

This equation states that one molecule of glucose plus six molecules of oxygen yield six molecules of carbon dioxide plus six molecules of water plus energy. The actual chemical process is more complicated than shown in the summary equation because there are many intermediate chemical reactions in the process. This process is described in more detail in later chapters.

Carbohydrates, fats, and proteins are the major energy-producing foods. The metabolism of fats and proteins is more complicated than that of carbohydrates because of their more complex molecular structure and chemical reactions.

Metabolism is influenced by such factors as species, race, age, sex, climate, season, and size, to mention a few (Becker and Hamalainen, 1914).

3.5.1.4. Classification of Animals. The system of classifying animals is the same as that used for plants. The present classification system for living systems was established by Carolus Linnaeus. This system provides an insight into the structure and characteristics of animals.

Determination of an animal's capacity to direct energy depends on those characteristics that differentiate one species from another, that define the energy relations with the environment, and that determine the internal energy of the animal. These characteristics include (1) mass, (2) temperature, (3) metabolism, (4) method of energy transfer between the environment and the animals, (5) method of extracting energy from the environment, (6) work done on the environment, and (7) structural organization.

3.5.2. Internal Energy

To obtain the capacity to direct energy of an individual animal, it is necessary to determine the internal energy of the individual and the subset of this energy that is directed by the individual. An animal's internal energy can be considered in terms of the constituent elements of the animal. The characteristics and types of the elements which constitute an animal are either known or can be determined. Therefore, the energy of an animal can be calculated by determining the energy level of each element and summing these energies.

The internal energies of an animal include mass, ground state electronic, chemical, mechanical, and electrical energies. The mass and ground state electronic energies are those associated with the elements and mineral molecules in an animal. These energies are the same types that were described above for minerals' and plants' elements and molecules. These kinds of energy for animals are like those for minerals and plants.

Animal cells carry out thousands of biochemical reactions per second—energy is either liberated or used in each of these reactions. The sum of these biochemical reactions is an animal's metabolism. The energies associated with these reactions are an element of an animal's internal energy. Some of these biochemical reactions are associated with the assimilation of food into protoplasm. During assimilation, energy in food is converted into chemical energy which is used in the synthesis of protoplasm. The chemical energies in the metabolism of glucose, as given in the above respiration equation, also are internal energies.

A significant difference between the internal energies of animals and plants is due to an animal's contractile tissue. The major characteristic of contractile tissue is its ability to convert chemical energy to mechanical energy and heat. An animal's contractile tissue produces the mechanical energies associated with heartbeat, blood circulation, breathing, and digestion. The energies associated with these mechanical functions are known and can be either measured or calculated. An animal's motion and locomotion is a function of its skeletal contractile tissue. The energies associated with skeletal muscle also can be measured or calculated.

Animals, like plants, can store foodstuffs internally, usually in the form of fats. The energies in these foodstuffs are internal energies.

The total energy in an animal includes a combination of the following:

- Mass energy
- Ground state electronic energy
- Electronic energies
- Chemical energies
- Mechanical energies
- Energies in stored foodstuffs

3.5.3. Environmental Energy

An animal's environment is much more restrictive than that of minerals or plants. Animals can only exist in a small portion of the environment where minerals can exist—minerals can exist in a range of high and low temperatures that would kill animals. Because animals feed on biochemicals that were initially synthesized by plants, they can only exist in environments where plants live or have lived. Both of the basic chemicals (glucose and oxygen) required for an animal's respiration metabolism are the byproducts of plants' photosynthetic process. These byproducts contain chemical energies essential for animal life.

An animal's biochemical reactions (metabolism) occur in very restricted thermal environments. The majority of animals can only live in thermal environments that range from the freezing to the boiling points of water. The thermal energies in an animal's environment are essential to maintain biochemical reactants at an energy level that promotes biochemical reactions. Both the chemical and thermal energies can be identified for the individuals of the various species, and these energies can be either measured or calculated.

The other energies in an animal's environment, such as radiant energies, are important to a lesser degree than those just mentioned. They can be identified for specific animals and can be measured or calculated.

3.5.4. Exchange Energy

Animals must continuously exchange matter and energy with their environments in order to live. Indeed, the definition of life includes the metabolic criterion, which requires a continuous exchange of matter and energy with the environment to carry out the chemical reactions of metabolism. Inputs to an animal from its environment include mass, chemical, heat, radiant, and mechanical energies. Outputs from an animal to its environment include the byproducts of metabolism. Also, animals exchange heat energies with their environments, just as plants and minerals do.

To determine the amount of directed energy exchanged between an animal and its environment, it is first necessary to identify the type and amount of energy that is exchanged. We must then determine the type and amount of exchanged energy that is directed by a individual animal.

Energy exchanges between animals and their environment are readily identified due to the robust literature on this subject, especially on human metabolism and the mass and energy exchanges between an animal and its environment that are necessary for metabolism. Animal exchange energies are identified below starting with those associated with metabolism.

A certain amount of energy must be received by an animal from its environment just to maintain basic organism functions. This minimum energy for maintenance is called *basal metabolism*. A generally accepted definition of basal metabolism for humans is the metabolic rate of an individual lying perfectly still, sufficiently long after a meal so that no digestion is taking place, and at a temperature range between 30° to 35°C. Metabolism is normally measured in calories per day. The effect on metabolism of changes in energy intake is illustrated by an experiment performed by Benedict *et al.* (1919) with a squad of athletic men. These men, whose normal daily intake was 3,200–3,600 calories, were placed on a diet containing 1,400 calories for a period of three weeks. The men lost, on an average, 12 percent of their weight and their basal metabolism was reduced 18 percent. The men were able to maintain this lower weight on 1,950 calories per day; however, on this reduced intake they were not as active as they were previously and had a lower tolerance of cold temperature.

Basal metabolism for humans is measured under thermal and other environmental conditions normal for humans. Basal metabolism for other animals can be measured for environmental conditions "normal" for them. In general, metabolism is the energy flow into an animal for its chemical reactions under a given set of conditions for the animal and the environment. These conditions usually include a temperature which is near "normal" for the animal whose metabolism is being measured. Temperature has a significant effect on metabolism, as illustrated by the work of Horst, Mendel, and Benedict on the metabolism of albino rats during prolonged fasting at environmental temperatures of 16°and 26°C. Both groups of rats had an average weight of 222 grams at the beginning of the fast. The total metabolism of the individual rats kept at 16°C was definitely higher (about 80 percent) than that of the animals kept at 26°C. The metabolic rate of the group in the 16°C environment was rather constant while that of the group in the 26°C environment had an average decline of 36 percent on the seventh day. As a result, the loss in weight was more rapid in the group in the 16°C environment. The animals fasting in the 26°C environment survived for an average of

16.5 days and lost 49 percent of their initial body weight. Likewise, the animals in the 16°C environment lost almost as much weight (44 percent) within a period of 11 days which was the average survival time of this group.

An animal's energy exchange also includes the passage of energy from the animal into its environment. Energies passing from an animal to its environment include mass, heat, and mechanical energy. The mass transferred to an animal's environment includes the byproducts of metabolism and the liquids used to cool warm-blooded animals. Radiant energy passes to the environment in the form of heat energy. Heat energy is also transferred to the environment by conduction and in the heat of the mass that is passed to the environment. Mechanical energy transferred to the environment is in the form of work performed on the environment.

The mass transferred by an animal to its environment as a result of metabolism includes carbon dioxide and water resulting from the respiration reaction and the byproducts of the other chemical reactions of metabolism. Also, this mass includes the waste products from digestive processes. There are energies associated with these various masses.

Heat energy transferred from an animal to its environment is a function of the surface area of the animal. The larger an animal's surface, the more heat energy the animal can transfer to its environment. Also, the amount of heat transferred to the environment is related to the animal's temperature and that of its environment. As an example, a larger amount of energy is required to maintain the body temperature (of warm-blooded animals) in extremely cold climates than in a warm climate. These heat transfer energies can be either measured or calculated.

Mechanical energies transferred by an animal to its environment are associated with the work performed by an animal on its environment. These mechanical energies are the result of the contractions of an animal's skeletal muscles. The energy exchange mechanism is the conversion of a muscle's chemical energy into mechanical energy associated with muscle contractions. Processes exist for determining the amounts of chemical and mechanical energies of muscle contractions.

3.5.5. Directed Internal Energy

Determination of an animal's capacity to direct energy was made by analyzing the internal energies identified above that are operated on in some way by animals. A summary of this analysis follows.

The energy in the mass of animals, as represented by the term mc^2 is not changed or used by an animal in any way. That is, none of an animal's mass is converted into atomic energy. This result of the analysis is due to the fact that the mass of animals is an assembly of mineral elements and the mass energy of these minerals is not converted to energy.

Like both mineral and plant individuals, animals cannot operate or change their ground state electronic internal energies. The rationale for minerals, plants, and animals is the same because of their inability to direct ground state electronic energies. That is, they all cease to exist if the ground state electronic energy is changed. Therefore, ground state electronic energies are not directed by animals.

The chemical and electronic energies associated with the chemical reactions of an animal's metabolism are operated on by the animal's structure. It is an animal's structure that allows these biochemical reactions to occur. For example, it is the animal's protoplasm in the form of enzymes that provides the catalysis that causes the biochemical reactions of metabolism. It was determined that the biochemical reactions of metabolism are operated on by animals, and that energies in these reactions are directed by animals. It is an animal's structure that provides it with a capacity to direct its metabolic energies.

An animal's mechanical energies are the result of behaviors of its contractile tissue. These mechanical energies are the result of an animal's muscle structures converting chemical energies to mechanical energies. That is, an animal's muscles operate on chemical energies and convert it to mechanical energies. Therefore, an animal has a capacity to direct mechanical energies.

Foodstuffs in the form of biochemical energies (e.g., fats) stored in an animal are operated on by an animal to convert these energies into forms that are readily used in chemical reactions. Therefore, these foodstuffs are operated on by an animal and are directed energies—that is, an animal has a capacity to direct these energies by virtue of its structure.

3.5.6. Directed Environmental Energy

All animals are capable of rejecting environmental energy at their surface in much the same manner as minerals and plants. The environmental energy rejected by the animal is directed environmental energy since the animal has operated upon this energy and changed its direction. The form of the rejected energy is much the same as that for minerals and plants. A large portion of the energy rejected by most animals is in the form of radiant energy. The capability to reject energy is a characteristic of the animal species.

3.5.7. Directed Exchange Energy

The energies accepted by an individual and those that are transferred from an individual to its environment have passed through the boundary of the individual. An animal's boundary operates on environmental energy to allow it to pass from the environment into the individual. This boundary structure also operates on an animal's internal energy to allow it to pass to the animal's environment. Therefore, the energies exchanged between an animal and its environment have been directed by the animal.

The amount of exchange energy an individual can direct is a function of the specific characteristics of this individual and especially of its boundaries. It is these characteristics that determine an animal's capacity to direct energy.

3.5.6. Total Capacity to Direct Energy

An animal's total capacity to direct energy is a combination of the internal, environmental, and exchange capacities to direct energy of the animal.

If an individual is in energy equilibrium with the environment (i.e., steady state), the internal energy of the animal, the directed energy in the environment, and the input and the output energy rates are constant. Under these steady-state conditions, the total capacity of the animal to direct energy is constant. A mature animal of constant weight on a steady regime of diet and work, that does not store food or other energy for use at a later time, is approximately in a steady-state condition. The maximum total capacity to direct energy occurs when (1) the energy input to the animal is as large as it can be, consistent with the physical limitations of the animal, (2) the energy output is the maximum work capability of the animal, (3) the internal energy is the maximum amount the animal can have and still function normally, and (4) the environmental energy is being directed consistent with the maximum work output exerted on the environment.

3.6. Summary

The behaviors of both living and nonliving systems can only be observed and quantified by the energies in these behaviors. These behavioral energies are different from the energy inputs to these substances. For example, an essential energy input for animals is food, but their major behavioral energies are observed to be heat and mechanical energy. It is obvious that the animals operate on, change, control, or otherwise direct energy. It is not obvious which energies in the environment were directed by substances, and this subject is discussed in detail in *A Measure of Knowledge*. Only summary results of research into substances' capacities to direct energy are provided in this chapter. However, these summary results are sufficient to demonstrate that all substances have a capacity to direct energy, which is a function of their structure and organization.

The major results of this research are:

- When energy takes form and becomes a substance, such as a mineral, a plant, or an animal, it acquires a fundamental behavioral characteristic which is a capacity to direct energy.
- A substance's capacity to direct energy can be specified and quantified.

CHAPTER 4

Behavioral Information

4.1. Introduction

Having proved that a system's capacity to direct energy is a fundamental behavioral determinant of both nonliving and living systems and that this determinant can be quantified, the question is: Are there additional fundamental determinants of behaviors? It was concluded that (1) available energy and a capacity to direct energy are the fundamental determinants of nonliving systems' and living systems' behaviors, and (2) there is an additional determinant of living systems' behaviors—information.

The nature, measurement, and measurement units of information have been established. The nature of information is much like that of energy—both are weightless and do not occupy space and therefore are abstract concepts. Both information and energy are associated with work—energy is the ability to do work while information is the ability to cause work. Energy is measured by the work it does whereas information is measured by the work it causes. Living systems' information can be classified as neural, chemical, or genetic.

To determine, what, if any, additional parameters are needed to quantify the behaviors, we consider the adequacy of extant parameters and their measures. Based on the evolution of the quantitative sciences, as described in Chapter 1, the existing quantitative sciences are based on the extant fundamental parameters. The capacity to direct energy, as described in Chapter 3, is a fundamental determinant of behavior that had not previously been identified. However, this measure is implicit in the structure and organization of nonliving systems. The existing literature considers structure and organization as fundamental to the behaviors of nonliving systems. The success of the existing quantitative sciences, which treat nonliving systems, is a powerful rationale for the adequacy of the existing determinants of nonliving systems behaviors. Also, the robust nonliving systems literature does not support there being any additional fundamental parameters.

On the other hand, the living systems literature identifies another determinant which is essential for behavior. That determinant is information. For example, Miller, in *Living Systems,* identifies information as being a major determinant both of living systems and of their behaviors. However, this information determinant does not have a measure equivalent to the fundamental measures of length, time, mass, temperature, charge, and energy. We propose that information is a fundamental determinant of behavior that can be measured just like the existing fundamental measures. This postulate is based on the model provided by the evolution of the quantitative sciences, as presented in Chapter 1. The development of this measure for information is provided here.

This chapter describes research leading to (1) the self-evident truth that the behaviors of living systems are a function of information, (2) proof that behavioral information is a fundamental determinant of behavior, (3) proof that behavioral information can be quantified, (4) the establishment of a measure and units of measure for behavioral information, (5) proof that behavioral information is a fundamental measure like energy, and (6) identification of the nature of information.

The quantitative sciences evolved from readily observed behaviors. Therefore, the quantification of information research was started based on the self-evident truth that our most easily observable behaviors, such as walking, running, and other movements, are functions of information. These behaviors are the result of muscle contractions. Our muscles must be "told" to contract in order for these easily observed behaviors to occur. We can inform muscles to contract, and thereby cause behaviors such as the movement of our hands, head, feet, or other body parts. This information is in the electrochemical neural impulses which cause muscle contractions. In addition, we can inform other people that we want them to behave in a particular way, and can observe that this desired behavior results. The neural information associated with electrochemical impulses and muscle contraction is treated first. After developing the measure for the neural information associated with muscle contraction, the behavioral information associated with the chemical behavior of living systems is presented. This information is called chemical information. This is followed by a treatment of the genetic information associated with the structure and organization of living systems, that is, the formation and maintenance of protoplasm.

4.2. Neural Behavioral Information

The most readily observed behaviors of animals are the rapidity of the motions of animals as compared to those of plants. The cause of these rapid motions in animals is the contraction of an animal's contractile tissue. It is the contraction of muscle tissue that causes the motions we can observe easily in animals. The muscle associated with these observed behaviors is composed of muscle fibers which are attached to a motoneuron. The readily observable behaviors (motions) of animals are the result of the contraction of the muscle fibers of one or a number of motor units. In addition to the readily observable behaviors of animals, there are behaviors that are more difficult to observe, such as the movement of food and food byproducts through the intestinal tract. These internal movements are also the result of the contractions of motor units.

Motor units are classified into two major types: those associated with an animal's skeleton and those associated with soft tissue, such as intestines. Skeletal motor units are normally composed of striated muscle fibers, and soft tissue motor units are normally composed of a different type of muscle fiber (smooth muscle). The observable behaviors of animals derive from the contraction of both types of motor units, with the exception of the few simple animals lacking defined motor units. Since easily observed animal behavior is related directly to the contractions of muscles, which are composed of motor units, the motor unit is a basic elemental behavior unit of animals.

The behavioral characteristics of motor units can be expressed in terms of the behavioral energy, available energy, and a motor unit's capacity to direct energy determinants treated in the preceding chapters. The behavioral energy associated with the contraction of a motor unit is in the form of work and heat. When the muscle fibers of a motor unit contract, they reduce their length and can perform work. They also generate heat during a contraction of the muscle fibers. This behavioral energy of contraction can be measured. The contraction of a motor unit's muscle fibers requires an input of available energy which is equal to the amount of work and heat energy resulting from the contraction. This is, of course, due to the conservation of energy law. The available energy used in the motor unit's contraction is in the form of chemical energy. The amount of behavioral energy and available energy associated with a motor unit's contraction is a function of the motor unit's structure and organization.

The relationship between information and muscle behavior is explored in detail by Simms (1991), but the fundamentals are explained here.

The existence of a relationship between a nerve impulse input to a muscle and the muscle's behavior is a well-established phenomenon. This phenomenon is demonstrated each time an output from the central nervous system, in the form of a nerve impulse to a muscle, activates muscle fibers causing these fibers to contract. The relationship between a nerve impulse and the resulting muscle fiber contraction is invariant. The invariant relationship is between the muscle nerve (motoneuron) impulse and the contraction of muscle fibers which always occurs in the healthy, non-fatigued muscle fibers associated with the motoneuron. The total energy produced by a muscle contraction can be measured and this energy characterizes the contraction behavior of the muscle. It was shown in Chapter 2 and previously (Simms, 1983) that behavior can be quantified by the types and amounts of energy associated with the behavior. The relationship between an information input to a muscle, in the form of a motoneuron impulse, and the energy produced by a contraction of muscle fibers is used to define and specify a unit of this neural information which causes muscle behavior.

A methodology for measuring the total energy produced by a muscle contraction is also well established. Total energy is the sum of the mechanical work and heat energy produced by a muscle each time the muscle is activated by an impulse in the muscle's motoneurons. This measurement methodology provides a means for quantifying the type of information which causes muscle contraction and for establishing a unit of measure for this information. This type of neural information can be called neural behavioral information because of its unique relationship to behavior.

The measure of neural information and its measurement unit are based on well-established phenomena, have arbitrarily selected units based on the phenomena, and are irreducible to other units. These measurement characteristics for neural information are the same as those for other fundamental measures, such as length, mass, time, temperature, electric charge, and energy. The measure of muscle behavior and its related neural information provides a means for quantifying the neural information in a single motor unit contraction, a series of neural information inputs to a motor unit, simultaneous neural information inputs to multiple motor units, and combinations of series and simultaneous neural information inputs.

4.2.1. Basic Concepts

The concept that muscle behavior can result when an input is received from an individual of our species, from another species, or from the environment is readily observable. Communication from other individuals of our species can cause an observed action, which is a manifestation of coordinated muscle contraction behavior. Inputs from other species in the form of aural or visual phenomena can cause coordinated muscle contraction behavior—such as running to avoid threats from other species. Also, visual and aural inputs from the physical environment can cause muscle contraction behavior in the form of retreat, investigation, or avoidance.

Measurement is based on a concept or a phenomenon which lends itself to measure. The concept of distance existed long before there was a method for its measure. Likewise, the concepts of time and mass existed prior to standards being established for comparing different time durations and masses. The concept that muscle contraction behavior is the result of an information input is the conceptual basis for measuring behavioral information.

Measurement is also based on establishing invariant standards (e.g., the centimeter, gram, second, and degree centigrade) against which comparisons can be made. Other standards are derived based on relationships among parameters and the invariability of these relationships. A heat energy input into a substance which causes a temperature increase is an example of a relational concept. The invariance of the amount of heat energy required to elevate one gram of water by one degree centigrade is the fundamental measure of heat energy and provides the quantitative standard for thermodynamics. Similarly an invariance also exists between a neural information input to a particular motor unit and the amount of energy produced by the resulting muscle contraction.

4.2.2. Information/Behavior Relationship

Communication of information from other individuals of our species, and inputs from other species and from the environment, act on (stimulate) the sense organs to produce electrochemical events in the nervous system. These stimuli may provide information from other individuals and about both nonliving and living systems in our environment. That is, the inputs contain information or patterns from which we obtain information. Following Galvani's discovery in the 1880s that an electrical stimulus causes frog legs to twitch, a direct relation between stimulus and muscle response was established. Galvani's discovery is described by Galambos (1962) as providing the basis for understanding the nerve impulse/muscle response relationship.

A motoneuron impulse is the final output of the master coordinating process of the central nervous system and is the result of neural information processing by the nervous system. The information processed by the nervous system may come from parts of the nervous system itself, from the environment via sensors, and/or a combination of these internal and environmental sources. The motoneuron is a very selective carrier of information. It will transmit only a message which directs the muscle fibers associated with the motoneuron to contract.

Although a motor unit consists of only one motoneuron, the number of muscle fibers in a motor unit may vary from one to hundreds, depending upon the type of muscle. An information input to a motoneuron causes the muscle fibers of the motor unit to contract, irrespective of the number of muscle fibers in the motor unit. Lucas (1917) has shown that the contraction of a muscle fiber is independent of the strength of the nerve impulse which evokes it. This phenomenon has since become known as the "All or None Law." This law is equally applicable to simple nervous systems and to the complex nervous systems of humans, described by Nauta and Feirtag (1986), which has two to three million nerves leaving the central nervous system to animate muscle fibers. A particular muscle may consist of many motor units which can contract simultaneously or in various combinations to produce specific behaviors.

Each time motoneuron impulses cause a muscle contraction, heat and work energies are produced by the muscle. Both the heat and work energies can be measured. Carlson *et al.* (1963) measured the total energy production (heat + work) in the contraction of the frog's sartorius muscle in response to nerve impulses that activated all the motor units in the muscle. The method used by Carlson *et al.* for measuring the energy in a sartorius muscle contraction is applicable to other skeletal muscles as well. Because of the differences in structure among various muscle types, the muscles each have different total energy outputs.

A methodology similar to that used to measure heat energy is applicable for measuring the behavioral information associated with muscle contraction behavior. The measurement of heat energy is based on the concept that heat energy inputs to a substance will cause an increase in the substance's temperature. The methodology is to select a standard substance—water—under specified conditions, then add heat energy to the water until its temperature is increased by a specified amount. This methodology was used to establish the standard measure for heat energy—the calorie. The heat energy characteristics of other substances are determined by comparison with the standard.

The methodology for measuring the behavioral information associated with muscle behavior is based on the concept that a neural information input to a muscle causes a contraction which produces work and heat energy. The muscle produces this energy by converting chemical energy into work and heat energies. The procedure is to select a standard motor unit, generate a neural information input to the motor unit, and measure the total energy produced by the motor unit. This standard measure for neural information can then be used for comparing the neural information input and energy output characteristics of other motor units.

A simple motor unit was selected as the standard for measuring neural information. This selection is comparable to selecting water, which is chemically very simple, as the heat energy standard. A comparable selection for the neural information standard is a muscle with a single motor unit. Unfortunately, measurements have not been made on the muscle contraction energy production of single motor unit. The motor unit of the complex sartorius muscle of Rana pipiens was used to develop the standard for measuring neural information because measurement data are available on the energy production of this muscle. However, these energy production data are for the whole muscle, and, for my purposes, must be converted to energy production for a single motor unit.

The energy production per motor unit was derived from the characteristics of Rana pipiens sartorius and the work of Carlson *et al.* The Rana pipiens sartorius is a skeletal muscle composed of motor units which Grinnell and Trussell (1983) describe as having a range of sizes. These size differences are due primarily to the number, rather than the size, of muscle fibers in each motor unit. The number of muscle fibers and their diameters are dependent on the size of the Rana pipiens. The number of fibers in mature Rana pipiens of 7.0–8.0 cm body lengths were determined by Yao and Weakly (1986) to be 804.0 ± 27.7 and the average diameter of the muscle fibers to be 58.0 ± 4.0 microns. The Yao and Weakly data are supported by measurements made by Sperry (1981).

Yao and Weakly counted 16.5 ± 0.9 motoneurons in the Rana pipiens sartorius using a chemical transport method. This number of motoneurons is supported by Grinnell and Trussell using an alternative method. They isolated 10 to 17 Rana pipiens sartorius motoneurons by gently teasing them away with glass microelectrodes or electrolytically sharpened tungsten wire tools. Since there is one motoneuron per motor unit, there are, using the Yao and Weakly data, an average of 16.5 motor units per sartorius muscle.

The number of motor units in a square centimeter of Rana pipiens sartorius muscle was calculated from the motor unit characteristics identified above. An average muscle fiber diameter of 58 microns was used to calculate the fiber's cross-sectional area. Assuming the fibers are circular in cross-section, the cross-sectional area of each fiber is 2,642 μm^2. Using an average of 804 fibers and 16.5 motoneurons per muscle, and assuming an equal distribution of fibers per motor unit, there are approximately 49 fibers per motor unit. The cross-sectional area of all 49 fibers in a motor unit is approximately 1.3×10^{-3} cm^2.

According to Saltin and Gollnick (1983), approximately one-third of the volume of muscle is made up of muscle fibers. Therefore, the cross-sectional area of muscle associated with a Rana pipiens sartorius motor unit is 3.9×10^{-3} cm^2.

Carlson *et al.* measured the total energy production of Rana pipiens paired sartorii muscle as a function of the load on the muscle. The data from these measurements were normalized to obtain the total energy production per unit mass of muscle, as shown in Figure 4.1. The abscissa in Figure 4.1 is the load P on the muscle as a fraction of the peak isometric (constant length) contraction tension P_{ot}. The left ordinate in Figure 4.1 is the total energy output of Rana pipiens paired sartorii muscle in millicalories per gram, as measured by Carlson *et al.* The measurement data presented in Figure 4.1 are the mean of 100 contractions, 20 by each of five muscles. Carlson and Wilkie (1974) stated that the density of muscle is approximately one gram per square centimeter of cross-sectional area. Therefore, the total energy production per gram of muscle or per square centimeter of muscle cross-sectional area is essentially the same. The left ordinate in Figure 4.1 is also the total energy output in millicalories per contraction per square centimeter of Rana pipiens sartorius cross-sectional area.

The total energy produced per square centimeter of sartorius muscle was converted to the energy produced in a single sartorius motor unit. Multiplying one mcal per cm^2 per contraction by the cross-sectional area of a motor unit (3.9×10^{-3} cm^2) obtains a conversion factor of 3.9×10^{-3} mcal (3.9 microcalories) per motor unit per contraction. This conversion factor was applied to the total energy production per gram of muscle, as measured by Carlson *et al.* to obtain the total energy produced per motor unit contraction as a function of load—shown by the right ordinate of Figure 4.1.

4.2.3. Unit of Neural Information

The total energy in microcalories per motor unit contraction, as a function of load, was used to establish a standard unit of measure for the behavioral information associated with muscle contraction. Establishment of the standard unit of behavioral information requires definitive specifications of the measurement characteristics:

1. The reference system is a motor unit of the sartorius muscle of Rana pipiens
2. The measurements are made under the conditions specified by Carlson *et al.* that is, the muscle tissue is at 0°C and a sufficient amount of chemical energy is available to the muscle so energy depletion is not a factor
3. The load on the muscle is a 0.45 fraction of the peak tension of a contraction, a value selected because it is near maximum energy production and it was a data point in the measurement of total energy production.

Using these specified conditions, a total energy production of 18.5 microcalories per motor unit per nerve impulse is obtained from Figure 4.1. This value is used to define the neural behavioral information unit as:

One unit of neural information causes the production of 18.5 microcalories of energy by a Rana pipiens sartorius motor unit at a

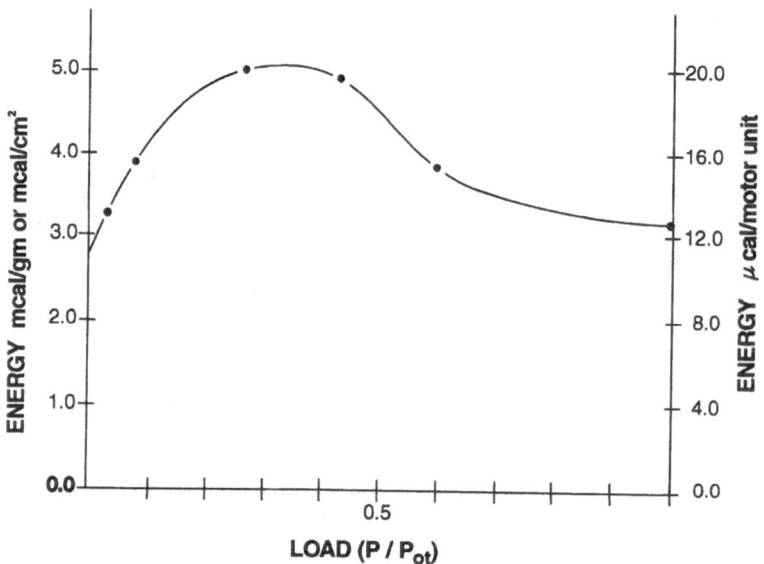

Figure 4.1. Energy per contraction vs. load.

temperature of 0°C and loaded to 0.45 of its peak tension load. This unit is named the neurin (a contraction of neural information.)

4.2.4. The Nature and Measurement of Neural Information

The information/behavior relationship and the development of a neural information unit of measure provides a basis for determining the nature of neural information. Major characteristics of neural information are the relationship between information, energy, and the measure of information in terms of energy. Therefore, the nature of information depends on the nature of energy. Because energy is weightless and does not occupy space, it is an abstract concept, as opposed to matter which has weight, occupies space, and is a concrete system. Neural information is also weightless and does not occupy space. Neural information, like energy, cannot be felt, touched, or held—it is an abstract concept. Energy is measured by its ability to *do* work. Because information causes the utilization of energy, it is measured by its ability to *cause* work. Another characteristic of neural information is that once it causes contractile tissue to contract, it ceases to exist. From these characteristics, it was determined that behavioral information is

- An abstract concept
- Weightless and does not occupy space
- Observed only by the energies used in living systems' behaviors
- Defined as that phenomenon which causes a living system's behavior
- Measured or calculated based on the behavioral energy it directs, i.e., the work it causes
- Ephemeral

The nature and measurement of neural information provide the basis for quantifying the effects of multiple information inputs to motor units and for information inputs to multiple motor units. It also provides a basis for making comparisons between various types of motor units.

The measurement of the information in Rana pipiens sartorius for various serial or parallel inputs is illustrated by the following calculations: A motor unit produces 18.5 microcalories for each unit of information input; therefore, the input of a series of 100 information units to a Rana pipiens sartorius motor unit results in an energy production of 1,850 microcalories. For a muscle consisting of 17 motor units, the simultaneous input of 17 units of information—one unit for each motor unit—causes the production of 17 times 18.5 microcalories or 315 microcalories for the complete muscle. The energy production of any combination of serial and/or parallel information inputs can be determined for a specified motor unit using these concepts of serial and parallel information inputs.

The measurement of information and energy in the motor units of other Rana pipiens muscles and in the motor units of other species requires the delineation of the characteristics of these other motor units. There are three significant behavioral characteristics of motor units: (1) They convert chemical energies into work and heat energies, (2) a specific motor unit has a unique capacity to convert energy, and (3) the conversion of chemical energy to work and heat energy results when a unit of information is present in the motoneuron.

A motor unit's capacity to convert chemical energy into work and heat energy is a function primarily of (1) the type of muscle fiber (Smith and Ovalle [1973] observed five fiber types in amphibian muscle), (2) the cross-sectional area of the fibers, and (3) the number of muscle fibers per motor unit. The type and size of the muscle fibers form the basic structure of a motor unit, and the number of fibers per motor unit is the organization of a motor unit.

A specific motor unit has a unique capacity to convert chemical energy to work and heat energy based on its structure and organization. A motor unit's unique capacity to direct energy can be used for comparison with the motor units of other muscles of the same species or of other species. The ability to compare motor units' capacities to direct energy provides a means for making comparisons to the Rana pipiens sartorius standard. The comparison of any motor unit with the standard is performed by applying one unit of information to the motor unit and measuring the total energy produced. The ratio of the total energy produced by the motor unit to the total energy produced by the Rana pipiens sartorius motor unit provides a comparison of the muscle's capacity to direct energy with the standard's capacity to direct energy.

There is a striking similarity between the ratio of a given motor unit's capacity to direct energy to that of Rana pipiens and the specific heat of substances. The specific heat of a substance is the ratio of the heat required to raise its temperature by 1°C with respect to the heat required to raise the temperature of water by that amount. The specific heat of a substance is one of the substance's fundamental characteristics just as the ratio of a motor unit's capacity to direct energy with respect to standard motor unit is a fundamental characteristic of a motor unit. Because of this similarity, the ratio of a motor unit's capacity to direct energy with respect to the capacity to direct energy of the standard Rana pipiens motor unit is named a motor unit's *specific capacity to direct energy.*

Given a motor unit's specific capacity to direct energy, its energy production for any specified information input can be determined. Also, if a muscle's specific capacity to direct energy and its number of motor units are known, the behavior of the muscle can be determined for any arrangement of information inputs.

4.3. Chemical Behavioral Information

The next most readily observable behaviors of animals after muscle contractions are those caused by biochemical reactions—an animal's metabolism. The investigation of these behaviors from a quantitative information point of view is described below. A brief summary of the findings are presented first, followed by a more detailed description.

The readily observable behaviors associated with an animal's metabolism include (1) ingestion of food, water, and air that are used as the reactants in chemical reactions, and (2) excrement, water, air, and heat eliminated as byproducts of chemical reactions. An animal's biochemical reactions are the basic cause of these observable behaviors. Therefore, biochemical reactions are the basic behaving elements of metabolism. Each biochemical reaction involves the rearrangement of chemical elements—the reacting molecules in a chemical reaction form other molecules and either require chemical energy or release chemical energy.

Biochemical reactions take place in animals because the animal's structures provide an environment in which biochemical reactions can take place. These structures and the biochemical reactants provide an animal with a capacity to direct the energies associated with biochemical reactions. Although an animal has a capacity to direct biochemical reactions, these reactions do not occur unless they are caused to do so by an enzyme. A particular enzyme molecule carries a single "message" for a specific behavior of rearranging the elements of molecules. From a behavioral point of view, an enzyme causes a biochemical reaction and, therefore, has the characteristics of information. The chemical information carried by enzymes is (1) an abstract concept, (2) weightless and does not occupy space, (3) observed only by the energies used in an animal's biochemical reactions, (4) defined as that catalytic action phenomenon which causes the biochemical reactions behaviors, (5) measured or calculated based on the behavioral energy it directs (i.e., the chemical energy it causes to be utilized), and (6) ephemeral because an enzyme must acquire another message before it can cause another biochemical reaction.

The concept of chemical information is not well understood. The literature describes hormones in terms of information, but enzymes are described as biochemical catalysts and not as chemical information transporters. Because the concept of chemical information is not understood, the detailed treatment of chemical information is introduced with a brief historical perspective of biochemical processes.

4.3.1. Historical Perspective*

Observation of biochemically caused behaviors can be traced to antiquity. The observed relation between the removal of an endocrine gland, such as castration performed as a religious rite, and changes in physical characteristics and behavior date far back in time. However, it was not until the work of Oliver and Schafer in 1894 that relationships between the various extracts of a gland and an animal's characteristics and behaviors were demonstrated. Oliver and Schafer administered various extracts of a gland and observed the changes caused by these extracts. In 1895, they showed that an extract of the suprarenal glands contained a substance "pressor," which, when injected, caused a large rise in arterial blood pressure. The active substance adrenaline (epinephrine) was isolated in 1901 by Takamine. A small amount of epinephrine injected into the bloodstream produces muscle constriction of almost all the small arterioles of the body, with a resulting rise in blood pressure. Its action is not confined to the blood vessels, but extends to all structures in the body that normally receive a nerve supply from the sympathetic nervous system. Thus, epinephrine increases the force of the heart and inhibits the movements of the digestive tract and, in many animals, of the bladder; it may also produce sweating, erection of hair, contraction of the pregnant uterus, and dilation of the pupil.

Oliver and Schafer showed, as in the case of the suprarenal gland, that an extract having a powerful pressor action (high blood pressure) could be obtained from the pituitary gland. The pressor action of the extract is due essentially to a direct action on the smooth muscle fibers of the arteries, causing them to contract. Indeed, in mammals the action of this pressor on the smooth muscle of practically all organs is to stimulate it to contract. For example, the pituitary extract has an intense stimulating action on the smooth muscle of the uterus.

The work of Bayliss and Starling in 1902 and 1903 (Bayliss, 1919) on the regulation of pancreatic secretion provided one of the best examples of the elucidation of the mode of action of a hormone. Hormone is the name they gave to the substances discharged from endocrine glands, a word derived from the Greek word for "I arouse to activity." They demonstrated that (1) secretin was a specific substance elaborated by the cells of the mucous membrane of the intestine, (2) the extracts from the mucous membrane of the small intestine of many animals caused pancreatic secretion when injected into a dog, and (3) the control of pancreatic secretion by the hormone secretin is a mechanism that exists in many vertebrates. The complete demonstration that secretin is present in the blood of an animal after the introduction of acid into the duodenum was given simultaneously in 1903 by Fleig and Enriquez and Hallion. The components of hormones were

*The pre 1950 history is based on the article "Hormones" in Encyclopaedia Britannica, Vol 11, 1951.

understood better as the result of the 1926 work of J. Mellanby, which suggested that the hormone secretin is a polypeptide.

By the early 1950s, the mechanisms of hormone action were beginning to be understood, due largely to the work of E.W. Sutherland (Sutherland, 1992). These mechanisms relate to the processes by which chemical messages are read by the cells that receive them. Two general mechanisms have been discovered, one for water-soluble hormones such as the peptide and protein hormones, and one for the lipid-soluble hormones—the steroids and thyroxine. The major differences between these two mechanisms is that the water-soluble hormones cannot penetrate a target cell's plasma membrane and must act through a second message, whereas the lipid-soluble hormones can readily penetrate the target cells' membrane and cause behaviors inside the cell. Water-soluble hormones act on the target cell by first binding to a receptor on this cell where the receptor passes the message through the plasma membrane and starts the generation of a second message internal to the cell. This second message is cAMP (cyclic adenosine $3',5'$-monophosphate) or other second messengers, such as 1,2-diacylglyceral. The characteristics of cAMP were first discovered by Sutherland, who was investigating how epinephrine stimulates liver cells to break down storage molecules and liberates glucose.

Several very important discoveries resulted from this research. It became evident early on that there were a number of steps between the hormone–receptor interaction and the liberation of glucose. Sutherland was able to show that one process involved was the control of enzyme activity through phosphorylation. This was the first demonstration of what is now known to be a common mechanism of regulation of enzyme function. Sutherland then demonstrated that epinephrine could stimulate fragmented liver cells to release glucose as long as pieces of their plasma membrane were present. This demonstration of hormone action in a cell-free system was also a major landmark in biochemistry. The third major discovery in this research program was that the interactions produced a small molecule that could then stimulate the phosphorylation of enzymes in a preparation of liver cell fragments that did not contain any membranes. This molecule was identified as cAMP.

Since Sutherland's work, identification of the number of systems activated by cAMP has grown rapidly. Many hormones in vertebrate tissue behave as a result of this second messenger. Different target cells have different specific secondary targets within them that are activated by the second messenger. These secondary targets can activate different behaviors in a cell. The specificity of hormone action resides not only in the receptors that determine which cells respond to a given hormone, but also in the way a given cell responds, and that depends on its responding mechanisms.

As stated above, lipid-soluble hormones have different mechanisms for causing behaviors in cells. Steroid hormones, such as estrogen, progesterone, and the hormones of the adrenal cortex, and thyroxine generally do not react with receptors on the target-cell surface. These hormones pass readily through the lipid-rich plasma membrane. They act by stimulating the synthesis of new kinds of proteins through gene activation rather than by altering the activities of proteins already present in the target cells. Once inside a cell, a lipid-soluble hormone binds to a receptor protein in the cytoplasm. The presence of a receptor protein is what distinguishes a responsive cell from a nonresponsive one. A receptor protein is specific for a particular hormone, and it changes shape when it binds its hormone. The hormone–receptor complex rapidly associates with acidic chromosomal proteins, and thus with the DNA of the chromosomes. The receptor protein itself cannot bind the acidic chromosomal proteins unless it has already bound a hormone molecule and undergone the necessary change in structure. Once associated with the acidic chromosomal proteins, the hormone activates the transcription of certain genes into messenger RNAs which are exported to the cytoplasm and translated into specific proteins.

The above historical perspective provides the relationships between the generation of chemical information (hormones) by glands and the resulting translation of materials and energy into specific proteins. However, metabolic behaviors occur in animals that do not have a hormonal system. These behaviors are the result of biochemical reactions that take place in single cells, as described below.

From a historical perspective, the details of these relationships in cells were observed only after the invention of high magnification instruments, such as the electron microscope, and new techniques, such as X-ray diffraction. The electron microscope was invented in the early 1930s, and the first attempt to apply the electron microscope to biological materials was in 1934. By the end of the period 1940–45, the electron microscope was firmly established as an extremely valuable tool for analyzing the structure of biological material and was used to identify the structure and organization of cells. The cellular structures important to chemical information are mitochondria, cytoplasm, and enzymes.

Mitochondria are organelles inside eukaryotic cells (cells that have a nucleus) whose behavior converts the energy in food materials into energy that can be used by the cell for biochemical reactions. This behavior is cellular respiration. It is the fundamental process of all living things, plants or animals, and is essential for the continuance of life. The chemical reaction for respiration has been known for many years and is expressed by the familiar formula:

$$C_6H_{12}O_6 + 6O_2 = 6CO_2 + 6H_2O + \text{energy} \qquad (4.1)$$

$$\underset{\text{glucose}}{} \quad \underset{\text{oxygen}}{} \quad \underset{\substack{\text{carbon} \\ \text{dioxide}}}{} \quad \underset{\text{water}}{}$$

The biochemical energy in this equation is in the form of ATP, which is stored energy that may be used either immediately or later to perform various kinds of work for the cell.

4.3.2. Basic Concepts

It was not until 1961 that biochemists and molecular biologists understood how respiration worked. In that year, the British biochemist Peter Mitchell proposed the chemosmotic theory which is a model that includes the relationship between structure and function. Mitchell proposed and then showed that operation of the respiratory chain results in the transport of hydrogen ions, against their concentration difference, through the inner membrane of the mitochondrion. Figure 4.2 is a cutaway view of a typical mitochondrion, showing its internal structure. The essence of the chemosmotic mechanism can be stated as follows: The flow of hydrogen through the respiratory chain results in a transfer of protons from the inside to the outside of the inner mitochondrial membrane, leading to an accumulation of protons on the outside. By the laws of diffusion, these excess protons tend to move back spontaneously into the region enclosed by the inner membrane (matrix), which they can only do by passing through the channel-like enzyme (ATP synthetase) molecules. In so doing, they provide the acidic conditions necessary for ATP production. Also, as protons diffuse away from an area of their high concentration, energy is released.

Discovery of the structure and organization of the organelle mitochondrion, as just described, provided the basis for stating that a cell's mitochondrial structure and organization provide the cell with a capacity to direct energy associated with respiration. Respiration is a capacity to convert food energy into chemical energies that can be used for the biochemical processes of the cell.

Biologists have known for a long time that the respiration chemical reaction expressed in the Eq. 4.1 does not go forward unless there are particular enzymes present. It is now known fact that for biochemical reactions to take place, an enzyme must be present. The mechanism of this reaction is that one enzyme mol-

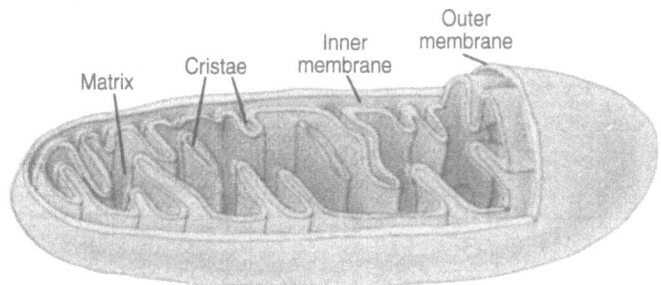

Figure 4.2. Mitochondrion (cutaway view).

ecule causes the biochemical reaction between the molecules of the reactants. This process is discussed in detail later.

4.3.3. Information/Behavior Relationship

The respiration biochemical reaction was selected for investigation of the fundamental relationship between behavior and information for a number of reasons. This reaction is a fundamental process of all living things, and has been extensively investigated.* The mitochondrion was selected because it is the elemental chemical behaving structure of eukaryotic cells; without this structure and its processes humans would not exist nor would the majority of animal species. Also, quantitative data exist which were used to determine the validity of conceptual information/behavioral relationships.

Respiration behavior of the eukaryotic cell begins in the cytosol (the fluid portion of the cytoplasm, excluding organelles and other solids), where chemicals from food substances become molecules that are then taken up by mitochondria. Figure 4.3 is a typical animal eukaryotic cell showing the parts of the cell. As stated above, mitochondria function primarily to capture energy from food substances in a form the cell can use. In mitochondria, energy-rich substances from the cytosol are oxidized, that is, electrons are removed from them. Some of the energy available from these electrons is used to make ATP that stores the energy in two special chemical bonds.

Typical mitochondria are approximately 1.5 micrometers in diameter and approximately 2 to 8 micrometers in length. Although mitochondria are visible with a light microscope, their structure was unknown until they were examined with the electron microscope. Electron micrographs show that they have an outer membrane that is smooth and unfolded. Immediately inside this is an inner membrane that folds inward at many points, giving it a much greater surface area than that of the outer membrane. In animal cells these folds tend to be quite regular, giving rise to shelf-like structures called cristae. The inner mitochondrial membrane contains large protein structures known to participate in cellular respiration. The region enclosed by the inner membrane is called the mitochondrial matrix and contains, among other things, some ribosomes and DNA that make some of the proteins needed for the synthesis of mitochondria.

Mitochondria are found in virtually all eukaryotes, except for a few microscopic forms that live in environments without oxygen. The number of mitochondria in a cell ranges from one contorted giant in some unicellular protists to a few hundred thousand in large egg cells. The average human liver cell contains

---

* The respiration chemical reaction process is described very well in current college-level textbooks—a detailed explanation of this process is provided in W. K. Purves, G. H. Orians, and H. C. Heller, Life: The Science of Biology, Sunderland, Mass.: Sinauer Associates, Inc. 1992.

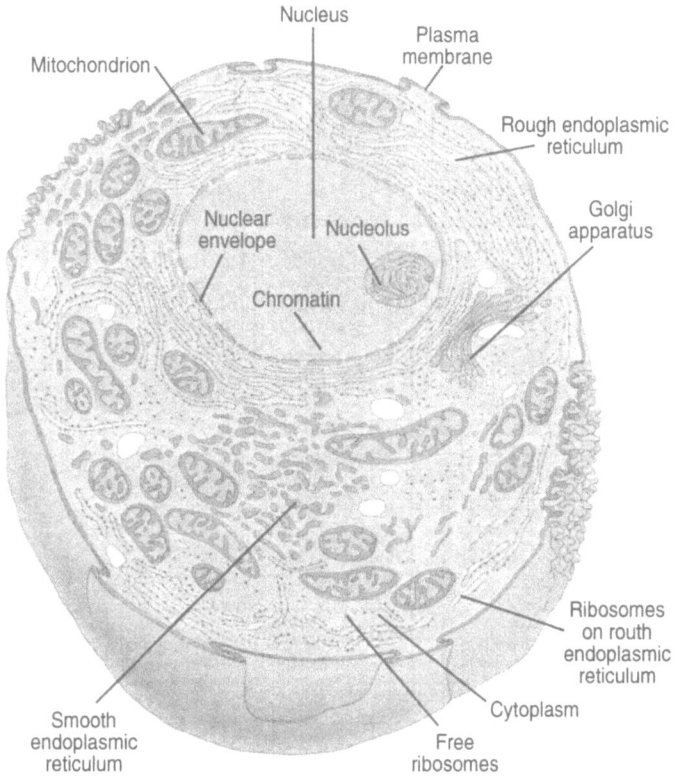

Figure 4.3. Typical animal cell (cutaway view).

more than a thousand mitochondria. Those cells that require the most chemical energy tend to have more mitochondria per unit volume.

The behavior of a cell in converting the energy in glucose to energy useful for biochemical reactions (as expressed in the Eq. 4.1) is a multi-step process that takes place in both the cytoplasm of a cell and in the mitochondria of the cell. Three processes, each consisting of a number of steps, are involved in the release of energy from glucose. The first process is called glycolysis (see Figure 4.4) and is performed within the cytoplasm of a cell. The second process (see Figure 4.5) is the citric acid cycle which takes place within the matrix of the mitochondria. The third process is the respiratory chain and takes place on the inner membrane of mitochondria. The second and third processes, together, constitute the cellular respiration process.

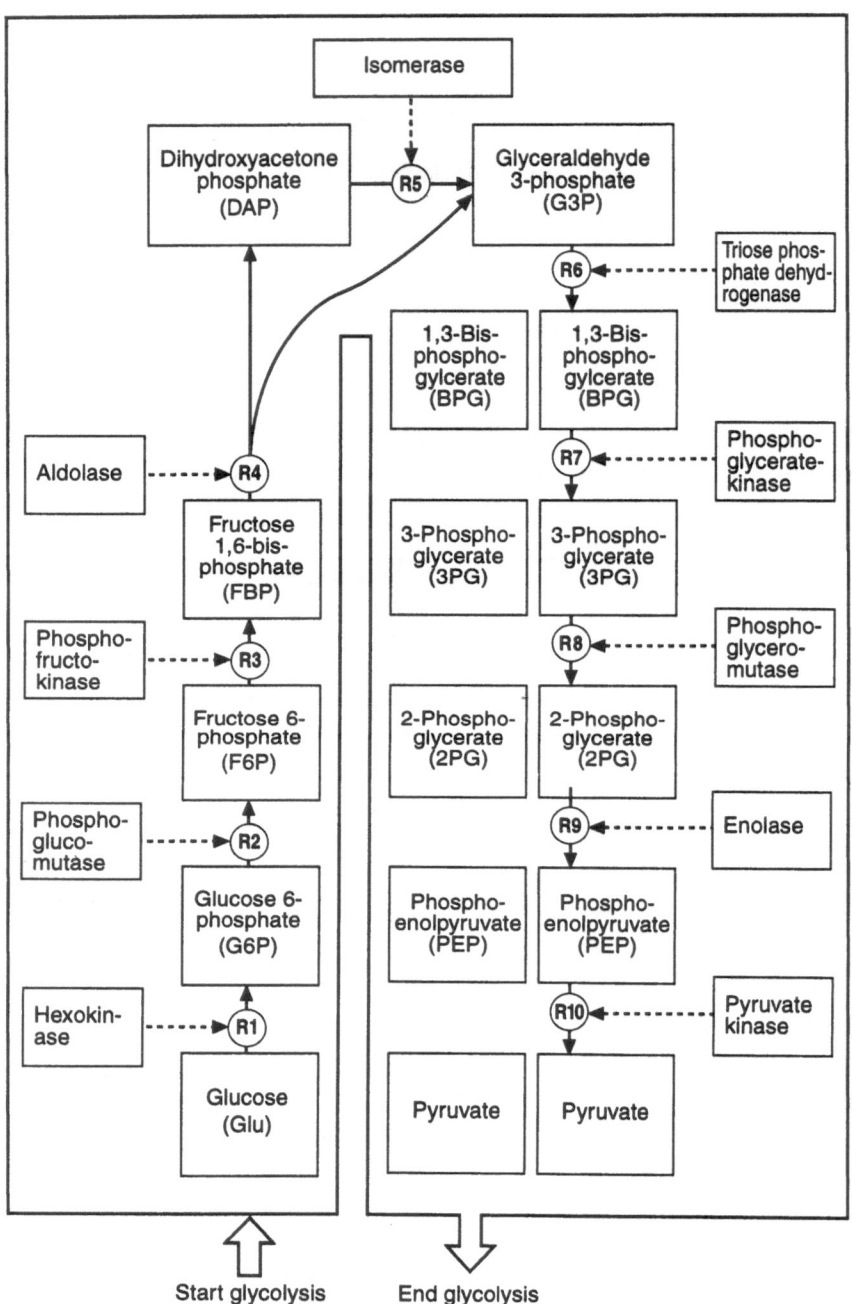

Figure 4.4. Glycolysis.

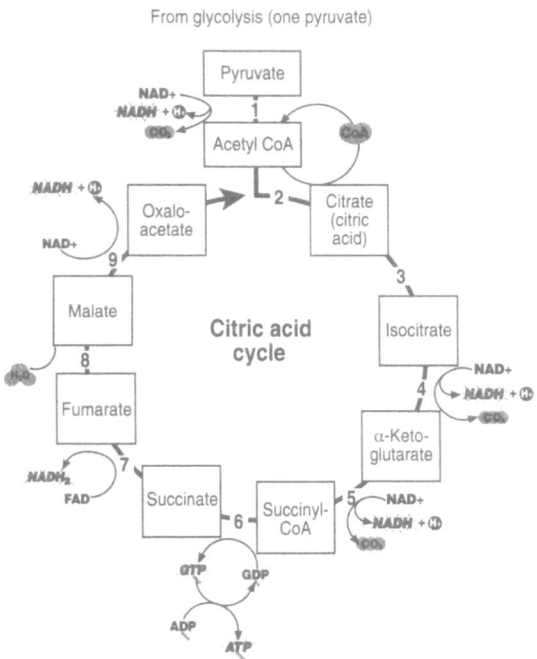

Figure 4.5. The citric acid cycle.(1) Pyruvate is oxidized to acetate, with the formation of NADH + H$^+$ and the release of CO_2; acetate is activated by combination with coenzyme A, yielding acetyl CoA; (2) the two-carbon acetyl group and four-carbon oxaloacetate combine, forming six-carbon citrate; (3) citrate rearranges forming its isomer, isocitrate; (4) isocitrate is oxidized to α-ketoglutarate, yielding NADH + H$^+$ and CO_2; (5) Alpha-ketoglutarate is oxidized to succinyl CoA, with the formation of NADH + H$^+$ and CO_2; the last carbon atoms of glucose are released in this step, which is almost identical to the first step; (6) succinyl CoA releases coenzyme A, becoming succinate; the energy thus released converts GDP to GTP, which in turn converts ADP to ATP; (7) succinate is oxidized to fumarate, with the formation of FADH$_2$; (8) fumarate and water react, forming malate; (9) malate is oxidized to oxaloacetate, with the formation of NADA + H$^+$. Oxaloacetate can now react with acetyl CoA to reenter the cycle.

4.3.3.1. Glycolysis. Virtually all living cells use glycolysis, the process by which glucose is metabolized to pyruvic acid (pyruvate). Glycolysis involves an oxidative step in which an electron carrier, NAD$^+$ (nicotinamide adenine dinucleotide), becomes reduced, acquiring electrons. In addition, a net yield of two molecules of ATP is obtained for each molecule of glucose that is processed through glycolysis. The major products of glycolysis are ATP, pyruvate, and the two electrons acquired by NAD.

In glycolysis, hydrogen atoms are acquired by the molecules they reduce. Most of the energy originally present as the covalent bonds of glucose is now associated with reduced NAD (NADH + H⁺). The respiratory chain is a process that releases that energy from the reduced NAD in such a way that it may be used to form ATP. This process is a series of successive reactions in which hydrogen atoms—or, in the later steps, electrons derived from hydrogen atoms—are passed from one type of membrane carrier to another and finally react with oxygen gas and produce water. In eukaryotes, these carriers (and the associated enzymes) are bound to the folds of the inner mitochondrial membranes (cristae). In both prokaryotes and eukaryotes, free energy drops with each transfer of electrons along the respiratory chain. The released energy is used to form ATP from ADP and P_i, where P is the phosphate ion HPO_4^{2-} and the i indicates an inorganic phosphate. This is the way in which the vast majority of the ATP in animals is formed. The formation of ATP during the operation of the respiratory chain is called oxidative phosphorylation.

As the energy is released, the reduced NAD and other agents of electron transfer are oxidized. They may then be reused in glycolysis and the citric acid cycle, steadily draining off hydrogen atoms and allowing those processes to continue to operate. This oxidation of NADH + H⁺ is thus another function of the respiratory chain. The inputs to the respiratory chain are stored hydrogen atoms and oxygen (O_2), and the outputs are water and stored energy in the form of ATP. The importance of ATP cannot be overstated. Without ATP cells cannot maintain their structure and metabolism and, as a result, the organism dies.

Respiration, as just discussed, can be described in terms of the fundamental principles of living systems behaviors. It is the structure of the cell, specifically the cytoplasm and the mitochondria, that provides the cell with a capacity to direct the energies associated with respiration. Food is the primary available energy, and ATP is the chemical energy that must be available for each of the endergonic reactions (those requiring an expenditure of energy) of respiration. Similarly, chemical information in the form of enzymes is necessary to cause the respiration behaviors, that is, the chemical reactions.

Quantification of information in terms of chemical energy requires a more detailed description of the respiratory process. To start with, the metabolic process leading up to respiratory metabolism consists of ten reactions of glycolysis, in which a molecule of the six-carbon sugar glucose is gradually converted into two molecules of the three-carbon compound pyruvic acid. These reactions (chemical behaviors) are accompanied by the net formation of two molecules of ATP and by the reduction of two molecules of NAD⁺ to two molecules of NADH + H⁺. That is, available energy is located in ATP, and four hydrogen atoms are passed on in a reducing agent. The further use of pyruvic acid depends on the kind of cell carrying out glycolysis and on whether the environment is aerobic or anaerobic. The use of NADH + H$^+$ is also variable. In most cases, it will be oxidized through the respiratory chain to yield water and NAD$^+$—a series of reactions that results in the formation of much more ATP (three molecules of ATP per molecule of NADH + H$^+$). In the fermentation process, NADH + H$^+$ is re-oxidized to NAD$^+$ either by pyruvic acid itself or by one of its metabolites, with no further storage of free energy. In either case, glycolysis may be regarded as a series of preparatory reactions, to be followed by either the citric acid cycle or the remainder of fermentation.

In the glycolytic process, shown in Figure 4.4, the first five reactions may be viewed as "pump-priming." Each of the five reactions require available energy. The behavior of the first reaction is to convert glucose to glucose 6-phosphate (G6P). The enzyme hexokinase provides the information for this reaction in which glucose receives a phosphate group from ATP. In the second reaction, glucose 6-phosphate is rearranged to form its isomer fructose 6-phosphate (F6P). The enzyme phosphoglucomutase causes this reaction. In the third reaction, fructose 6-phosphate (F6P) is converted to fructose 1,6-bisphosphate (FBP). This reaction is caused by the enzyme phosphofructokinase, which uses a phosphate from another ATP. In the fourth reaction, the enzyme aldolase causes fructose 1,6-bisphosphate to break into two different three-carbon sugar phosphates, dihydroxyacetone phosphate (DAP) and glyceraldehyde 3-phosphate (G3P). In the fifth reaction, the enzyme isomerase causes the DAP molecule to rearrange into its isomer G3P. By the end of these five reaction processes, two molecules of ATP (available chemical energy) have been used in the five reactions, and the glucose molecule has been converted into two molecules of a three-carbon sugar phosphate (G3P).

The behaviors of the second five reactions of the glycolytic process are to convert the chemical bond energy of each molecule of three-carbon sugar phosphate formed in the first five reactions into chemical energy (ATP) and pyruvate. These second five reactions are each mostly exergonic, that is, they form ATP. In the sixth reaction, the enzyme triose phosphate dehydrogenase causes a molecule of a three-carbon sugar phosphate (G3P) to gain phosphate groups and to be oxidized, thus forming two molecules of NADH + H$^+$ and two molecules of 1,3-bisphosphoglycerate (BPG). In the seventh reaction the enzyme phosphoglyceratekinase causes the two molecules of BPG to transfer phosphate groups to ADP to form two ATPs and two molecules of 3-phosphoglycerate (3PG). In the eighth reaction, the enzyme phosphoglyceromutase causes the phosphate groups on the two 3PGs to move, forming two 2-phosphoglycerates (2PG). The ninth reaction consists of the enzyme enolase causing the two molecules of 2PG to lose water, forming two high-energy phosphoenolpyruvates (PEP). Finally, in the tenth reaction, the enzyme pyruvate kinase causes the two PEPs to transfer their phosphates to ADP, forming two ATPs and two molecules of pyruvate.

In summary, glycolysis consists of the information contained in ten specific enzymes causing ten chemical reactions which convert glucose into pyruvic acid—2 mols* for each mole† of glucose that entered glycolysis. Further, at the beginning of glycolysis, two molecules of ATP are used per molecule of glucose, but ultimately four are produced (two for each of the two 1,3-bisphosphoglycerates)—a net gain of two ATP molecules and two NADH + H$^+$, with over 20 kcal of free energy stored in ATP for every mole of 1,3-bisphosphoglycerate that is broken down. The metabolism of glucose to pyruvate is accompanied by a drop in free energy of about 140 kcal/mole. About one third of this energy is captured in the formation of ATP and reduced NAD. Further free energy for biological purposes can be gained by oxidizing the pyruvate.

4.3.3.2. *The Citric Acid Cycle.* The citric acid cycle (also called the Krebs cycle or the tricarboxylic acid cycle) is a process performed in the liquid matrix of a cell's mitochondria. This process starts with pyruvate from the glycolysis process and is depicted in Figure 4.5. In the presence of oxygen, the pyruvate from glycolysis is oxidized in the citric acid cycle through a series of small chemical reactions. Each of these small steps are caused by enzymes present in the liquid matrix of the mitochondria. There is a specific enzyme for each of these steps. Various steps release the carbon atoms of pyruvate (originally the carbon atoms of glucose) as carbon dioxide (CO_2) molecules and transfer more electrons to carriers. The products of the citric acid cycle are carbon dioxide and many more stored electrons (along with accompanying hydrogen nuclei) than are produced in glycolysis—more stored electrons means a greater ultimate production of ATP.

The citric acid cycle takes pyruvate and breaks it down to CO_2, using the hydrogen atoms to reduce carrier molecules and to pass chemical free energy to those carriers. The reduced carriers are later oxidized in the respiratory chain. The principal inputs to the citric acid cycle are pyruvic acid, water, and oxidized electron carriers; the principal outputs are carbon dioxide and reduced electron carriers.

The citric acid cycle consists of nine reactions. In the first step, pyruvate is oxidized (yielding useful free energy) and converted to an activated form of acetic acid (CH_3COOH) called acetyl coenzyme A. Acetyl coenzyme A, with two carbon atoms in its acetate group, then reacts with a four-carbon acid (oxaloacetate) to form the six-carbon compound citric acid (citrate). The remainder of the cycle consists of a series of enzyme-catalyzed reactions in which citric acid is

---

* A mol (also referred to occasionally as pressure or volume fractions) is the ratio of a particular constituent in a gas compound to the total of all gases present—where both the constituents and the total gas are expressed in moles.

† A mole is a quantity of a compound whose weight in grams is numerically equal to its molecular weight expressed in atomic mass units (a.m.u.), where molecular weight is the sum of the atomic weights of the atoms in a molecule and atomic weight is the average weight of an atom of an element on the a.m.u. scale. For example a mole of glucose ($C_6H_{12}O_6$) is [(12.01 × 6) + (1008 × 12) + (16 × 6) = [72.06 + 12.096 + 96] or 180.156g.

degraded, leading to the release of two of the carbons as CO_2, to the production of useful free energy from redox reactions (chemical reactions in which one reactant becomes oxidized and the other becomes reduced), and to the production of a new four-carbon molecule of oxaloacetate from the other four carbons. The new oxaloacetate can react with a second acetyl coenzyme A, producing a second molecule of citrate, and so forth. Acetyl coenzyme A enters into the cycle from pyruvate, CO_2 is going out, the rest of the compounds in the cycle are being used and replaced, and energy from redox reactions is being stored. During the citric acid cycle as a whole, starting with a single molecule of pyruvate, three carbons are removed as CO_2, and five pairs of hydrogen atoms per pyruvate molecule are used to reduce carrier molecules, with the simultaneous storage of energy. The energy-removing reactions(see steps 1, 4, 5, 7, and 9 in Fig. 4.5) are the major function of the cycle. The amount of free-energy change in the cycle is more than 600 kilocalories per mole.

4.3.3.3. The Respiratory Chain. As indicated previously, the principal role of the respiratory chain is to release the energy from reduced NAD in such a way that it can be used to form ATP. The major elements for implementing this role are described below.

The inputs to the respiratory chain are stored hydrogen atoms and oxygen; the outputs are water and stored energy in the form of ATP. The primary chain consists of electrons passed through four electron transport complexes. The first reaction converts NADH + H$^+$ to a specific oxidizing agent called ubiquinone (Q)—this reaction is caused by the enzyme NADH-Q reductase, which is a complex of 25 polypeptide subunits. The second reaction is caused by the enzyme cytochrome reductase (a complex with nine subunits), which converts ubiquinone into cytochrome c. The third reaction is caused by the enzyme cytochrome oxidase (with eight subunits) which convert cytochrome c into H_2O. The last reaction is caused by the enzyme ATP synthetase, which converts ADP and protons to ATP.

The complete respiratory chain consists of the primary chain just described and a source of electrons provided by the succinate-to-fumarate reaction of the citric acid cycle. The enzyme succinate-Q reductase causes this input from the citric acid cycle to be converted to ubiquinone. For every succinate oxidized by the respiratory chain, two molecules of ATP are eventually produced.

4.3.4. Behavioral Energy

The glucose metabolic behavior of cells just described provides the data necessary for these behaviors to be described in terms of their associated energies. The amount of energy associated with glycolysis and cellular respiration has been determined previously by biologists.

Briefly stated, glycolysis yields two molecules of ATP for every glucose molecule entering the process. The ensuing citric acid cycle and respiration chain produce an additional 34 ATP molecules for every glucose molecule. The source of most of these ATP molecules is the oxidation of reduced carriers (produced in glycolysis and the citric acid cycle) by the respiratory chain. A yield of three molecules of ATP is obtained for each NAD^+ regenerated by the respiratory chain and two molecules of ATP for each FAD. Thus, the total gross yield of ATP from one molecule of glucose taken through glycolysis and respiration is 38. However, two ATP have been used in glycolysis to drive endergonic reactions, for a net yield of 36 ATP. This is because the inner mitochondrial membrane is impermeable to NADH, and one ATP must be used for each NADH (produced in glycolysis) that is shuttled into the mitochondrial matrix.

4.3.5. Capacity to Direct Chemical Energy

The data just presented are sufficient to conclude that it is the structure and organization of cells and their organelles that provide cells with a capacity to direct energies associated with glucose metabolism behaviors. This phenomenon of the relationship between structure and the behavior of the respiratory chain behavior has been known since Peter Mitchell proposed the chemosmotic theory in 1961.

4.3.6. The Nature and Measurement of Biochemical Information

The above data on glucose metabolism behavior also illustrate the fact that biochemical reactions are caused by enzymes. From the quantitative living systems point of view, it is chemical information that causes these biochemical reactions—just as neural information causes the behaviors associated with the contraction of motor units. Thus, enzymes carry the chemical information that causes biochemical reactions/behaviors.

4.3.7. Unit of Biochemical Information

A measurement unit for biochemical information was developed using the model for developing the unit for heat energy and following the techniques used for developing the measurement unit for neural information.

Selection of a reference biochemical reaction for the development of a biochemical information unit was based on the following criteria: (1) the reaction occurs in virtually all organisms, (2) the biological work required by the reaction is provided by a single molecule of ATP, (3) the enzyme that causes the reaction is known, (4) the amount of chemical energy required by the reaction is known, and (5) sufficient data are available to describe the reaction and its behavior. The free-energy changes for glycolysis are shown in Figure 4.6. The first reaction in the glycolysis process meets these criteria and was selected as the reference biochemical behaving element. A number of other reactions could have been selected because many different enzymes can cause the release of free energy from ATP. However, the glycolysis reactions are perhaps better known.

The first biochemical reaction of glycolysis is the conversion of glucose to glucose 6-phosphate by the transfer of phosphate from ATP to the six-carbon sugar glucose. ATP provides approximately 12 kilocalories of free energy per mole to cause this reaction. Conversion from energy per mole to energy per molecule, using Avogadro's constant, yields approximately 2×10^{-20} calories of free energy per molecule of ATP. This is the amount of energy caused to be utilized in the biochemical reference reaction by one quantum of biochemical information. Thus, one quantum of biochemical information causes the utilization of 2×10^{-20} calories of energy in the reference behaving system.

This biochemical information unit is approximate because the reference biochemical reaction system and its environment have not been completely specified. This unit is named the biocin (a contraction of biochemical information). This lack of specificity introduces uncertainties in the amount of free energy provided by ATP. The actual amount of energy provided varies as a function of the concentrations of ATP, ADP, and phosphate ion. The amount of energy is also a function of temperature and pH. The biochemical information unit can be made

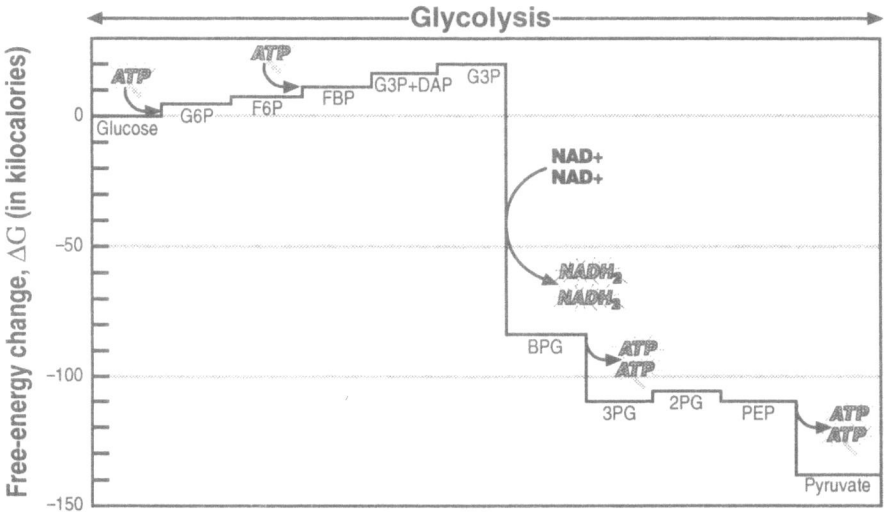

Figure 4.6. Free energy change during glycolysis.

precise by specifying the temperature, pH, and concentrations of ATP, ADP, and phosphate ion that result in exactly 2×10^{-20} calories per quantum of biochemical information.

4.4. Genetic Information

The readily observable behaviors associated with an animal's muscle contraction and metabolism can be observed over periods of time as short as minutes and hours. Another readily observable behavior is growth, which must be observed over longer periods of time such as weeks or months. These slow observable growth behaviors include changes in an animal's size, shape, and mass. Growth behaviors are the result of protein synthesis by an animal's cells. At a more basic level, growth behavior is the result of amino acids being synthesized into protein by an animal's cells. The investigation of growth behavior resulted in the following findings which are summarized first, then followed by a more detailed treatment.

Within an animal's cell structure are small organelles called ribosomes that are the site of protein synthesis. These ribosomes provide an animal's cell with a capacity to direct the energy necessary for protein synthesis. That is, the ribosome is a cell's basic behaving element for the synthesis of protein.

Although an animal has a capacity to direct energy for protein synthesis behavior, this behavior does not occur until genetic information causes it. Genetic information is associated with the formation of the structure and organization of animals. It determines how matter will be organized into structures that have specific capacities to direct energy.

A brief history of the evolution of genetics and genetic information is given below to provide background on the development of a behavioral point of view for genetic information.

The structure and organization characteristics of living systems as they are influenced by their environments and by heredity were studied by Charles Darwin. His book *Origin of Species*, published in 1859, described his findings and his theories on how species evolve. From a behavioral information point of view, Darwin's work is related to the adaptation of a living system's structure and organization to environmental changes, but it does not adequately address the importance of information in this process.

The concept of genetic information had its origins in the work of Abbot J. G. Mendel relating to the distributive mechanism of organic inheritance, promulgated in 1866. The essence of Mendel's research, as it applies to genetic information, is that characters exist in parent plants that are inherited by the following generation, causing unique structures and organization. In other words, there are factors passed from one generation to another that determine the structure and organization of the latter generation. Mendel had discovered the discrete units of heredity—the genes. His work was essentially unread until his laws of genetics were independently rediscovered at the beginning of the twentieth century. Although Darwin was working during Mendel's time (he published *The Descent of Man* in 1871) he was unaware of Mendel's work; the knowledge of Mendel's work would have made his task much easier.

Discovery of DNA's structure was a major step forward in understanding heredity and genetic information. Although nucleic acids were discovered in cells in 1868, their relevance to heredity was first suspected only in the 1940s. The structure of DNA, and the implication of this double helix structure to heredity, was first discovered in 1953 by James Watson and Francis Crick (1953). Their discovery changed the nature of genetics and provided an understanding of genetic information.

A robust genetic science literature has been developed since Watson and Crick's discovery. It is now known that a sequence of genes comprise the long, self-replicating, double-stranded molecule. The two strands, wrapped around each other make up the DNA double helix. The nucleotide bases in DNA are called adenine, cytosine, guanine, and thymine, which are abbreviated A, C, G, and T. Some genes may be composed of more than a million A's, C's, G's, and T's. Typical genes that control a specific hereditary trait are several hundred to thousands of nucleotides long. Also, it is now known that, except for a few microbes, the genetic information of every living system on earth is contained in DNA. A specific sequence of A's, C's, G's, and T's contains all the information necessary for making a living individual—one for humans, another for horses, another for birds, and so forth.

4.4.1. Basic Concept

Genetic information causes the production of specific proteins. This information is encoded in DNA, the genetic material, which consists of two polynucleotide chains forming a double helix. The central dogma of molecular biology is that DNA codes for RNA while RNA codes for protein. Figure 4.7 is a simple schematic of this central dogma. This figure also shows that biochemical information causes the replication and translation behaviors.

Like neural and biochemical information, genetic information is weightless and does not occupy space. Also, like neural and biochemical information, genetic information is generated in one location and then transmitted to another location where it causes a particular behavior. Genetic information generated by DNA causes the synthesis of protein. This DNA information is transcribed into two types of RNA that travel to the cytoplasm. One is a messenger RNA, the other is a transfer RNA which recognizes the genetic message and simultaneously carries specific amino acids, thus translating the language of DNA into the language of proteins. Genetic information is also like neural and biochemical information in that it cannot be directly measured, but it must be observed and measured by way of its behavior—the production of proteins.

The production of protein requires an expenditure of chemical energy stored in molecules of ATP. The amount of ATP required for the production of specific proteins can be calculated and represents the energy required for the protein production behavior. The production of protein also requires biochemical reactions

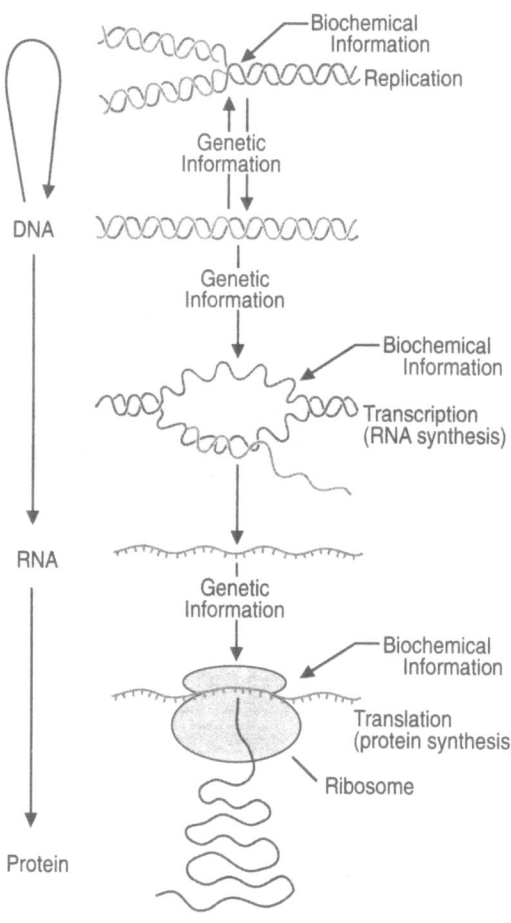

Figure 4.7. The central dogma.

that are caused by specific enzymes (biochemical information). The mechanisms for the production of protein are known, and the chemical reactions, energies, biochemical information, and genetic information have been identified (Purves *et al.*, 1992).

The facts and considerations given above demonstrate that genetic information has the same general characteristics as neural and chemical information.

4.4.2. Unit of Genetic Information

A measurement unit was developed for genetic information based on the model used for the development of a unit of measure for neural information. Selection of a reference protein synthesis system for the development of a genetic information unit was based on the following criteria: (1) the protein synthesized is simple, consisting of two linked amino acids, (2) the enzymes that cause the biochemical reactions are known, (3) the biological work required for synthesis is known, and (4) sufficient data are available to precisely describe the reference protein synthesis system and its processes. A reference protein synthesis system that produces an end product protein consisting of the amino acids methionine (met) and proline (pro) meets these criteria. The end product is the dipeptide protein met–pro that floats in the cytoplasm.

The reference protein synthesis system consists of (1) the DNA molecule; (2) the mechanism for converting the genetic information in DNA to genetic information in RNA; (3) two types of RNA, one a specific type of RNA molecule that is a complementary copy of one strand of the gene, called a messenger RNA or mRNA, and the other a specific transfer RNA or tRNA; (4) enzymes (aminoacyl-tRNA synthetases); and (5) ribosome consisting of two subunits, a larger, or heavy one and a smaller or light one. This system uses one DNA strand—the template strand—to transcribe genetic information in DNA into RNA information. The genetic information in RNA resides in mRNA and in tRNA. The system uses biochemical information in the form of enzymes to perform the biochemical reactions of converting from one form of genetic information to another, and of binding methionine and proline amino acids to their respective tRNAs to form charged tRNA molecules. The system's transmission medium transports these charged tRNA molecules to a cell's ribosome where they are operated on to synthesize protein.

The unit of measure for genetic information was derived from the work this information causes in the synthesis of the reference methionine–proline protein. Protein synthesis is a complex process; a simplified process is used to describe the reference protein synthesis system. The simplified process is to (1) identify the gene that contains the genetic information (i.e., codes) for synthesis of the methionine–proline protein, (2) identify how this genetic information is transmitted to the synthesis site (the ribosome) and the form it takes in the transmission, (3) identify the biochemical reactions necessary to synthesize the reference protein, (4) identify the biochemical energy and the amount of this energy necessary to perform the work to synthesize the reference protein, and (5) relate the genetic information to the total work.

The reference system gene consists of three codons—one for the process initiator methionine, one for proline, and one to terminate the synthesis process. Genetic information is encoded in codons (three-letter words) comprising the bases uracil (U), cytosine (C), adenine (A), and guanine (G). For example, the codon for methionine is AUG, that for proline is CCG, and that for termination is UAA. Transmission of the reference gene is initiated by the generation of an mRNA and two tRNA molecules. These molecules travel from the nucleus to the cytoplasm of a cell. The genetic information is now in a form that can cause synthesis of the reference protein.

Synthesis begins with biochemical reactions that charge the two tRNAs: a charged tRNA has been combined with the specific amino acid it is coded for and its energy level has been increased by ATP. Enzymes cause the charging of tRNA (See Figure 4.8). The ATP molecule in each reaction is converted to AMP and releases approximately 24 kilocalories per mole of free energy. This energy provides the biological work necessary to charge a tRNA.

Synthesis is completed in a ribosome structure where the genetic information encoded in the mRNA is interpreted by the charged tRNA molecules. The energy in the charged tRNAs is used to perform the work necessary to combine the two amino acids to synthesize the reference protein. Synthesis of the reference protein is described in Chapter 10 (Fig 10.2).

The genetic information in the reference system causes the work in the ribosome necessary to synthesize the dipeptide methionine–proline. This work is derived from the energy added to the tRNAs from the ATP that charges them. Assuming that half of the free energy from ATP is used to combine the tRNA molecule and the amino acid molecule, the other half (12 kilocalories per mole) is used to increase the energy level of the charged tRNA. The genetic information in the gene causes 24 kilocalories per mole of work in synthesizing the reference protein—12 kilocalories per mole from each of the two charged tRNAs. Under these conditions, the genetic information for the reference system causes 24 kilocalories per mole of work in the synthesis of the reference protein.

Because a gene consisting of three codons is the smallest size gene that can synthesize a dipeptide protein, the information in a three-codon gene is considered to be a quantum of genetic information. Thus, one quantum of genetic information causes 24 kilocalories per mole of work to synthesize the reference protein. A quantum of genetic information causes 4×10^{-20} calories of work to synthesize one molecule of the reference protein. This approximate genetic information unit can be made precise with specificity of the reference system and accurate measurement of the energies used in protein synthesis. This unit is named the genin (a contraction of genetic information).

4.5. Summary

It was postulated that information is a fundamental determinant of the behaviors of living systems and that a measure for this determinant could be developed. It was further postulated that this measure could be developed in the same way that the fundamental measures of length, time, mass, temperature, charge, and energy were developed. These postulates were tested and found to be true.

This fundamental determinant was named behavioral information in order to segregate it from other meanings of the term information. It was found that

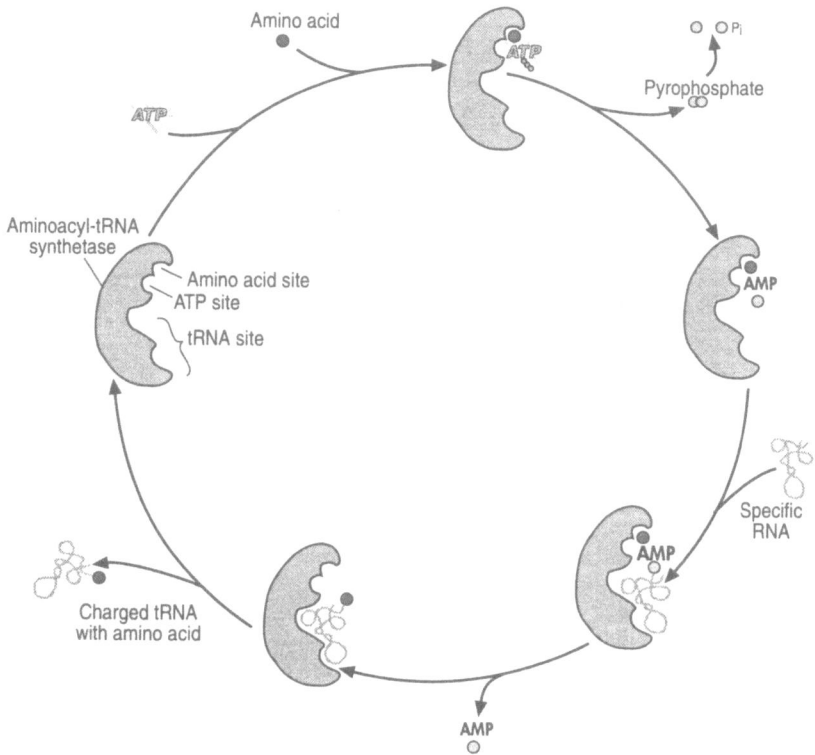

Figure 4.8. Charging a transfer RNA molecule.

behavioral information is a fundamental measure just as length, time, mass, temperature, charge, and energy. Units of measure were established for neural, chemical, and genetic information that are equivalent to units for the other fundamental measures.

It was determined that information is:

- An abstract concept
- Weightless and does not occupy space
- Observed only by the energies used in living systems' behaviors
- Measured or calculated based on the behavioral energy it directs (i.e., the work it causes)
- Ephemeral

CHAPTER 5

Fundamental Equations for the Behaviors of Animals

5.1. Introduction

Having identified the fundamental determinants of living systems' behaviors, we now develop the stages of a quantitative science evolution model. There are two stages: (1) the establishment of relationships among fundamental determinants, and (2) the development of equations that quantify the relationships. This requires a suitable model.

The quantitative science evolution model described in Chapter 1 was adequate for determining the fundamental determinants of behavior. However, it needed to be further defined and specified to be used as a model for establishing relationships among these determinants. This increased specificity was based on the extant quantitative sciences, which are for nonliving systems.

The more specific model was then used to develop the relationships and equations for the behaviors of individual animals. In this chapter, the behaviors of animals are limited to individuals; group behavior is the subject of a second volume.

5.2. The Quantitative Science Evolution Model

Specification of the evolution of a quantitative science model starts with geometry and proceeds through the electrical sciences. The subject of geometry is shapes. The fundamental determinant of geometry was shown in Chapter 1 to be length. Shapes are described in terms of length and the relative lengths and positions of the segments of these shapes—that is, the relationships among the segments of length of the shapes. For example, a square is defined in terms of the relationship between four straight lines of equal length, and a rectangle in terms of the relationship between four straight lines with two lines of one length and two lines of another length. Also, there is a relationship between the area of a shape and the lengths of the shape's segments. Equations were developed long ago for the relationship between, for example, area as a function of the length of segments of geometric shapes. These simple examples provide an indication of the evolution of a quantitative science from determinants to relationships to equations.

The first-order determinants of astronomy are length and time, as described in Chapter 1. Kepler formulated a relationship among the positions of celestial bodies in terms of distance from the sun and of time. This relationship is given in Chapter 1 along with the equations which are known as Kepler's laws. This demonstrates the evolutionary stages of the quantitative sciences which proceed from identification of determinants, to development of relationships among determinants, and then to the development of quantitative equations. The individual subjects of astronomy are the celestial bodies.

The first-order determinants of mechanics are length, time, and mass, also described in Chapter 1. Newton discovered relationships among these determinants as they apply to physical bodies. One relationship, known as Newton's first law of motion, states that each and every body continues in a state of rest or uniform motion in a straight line, unless it is compelled to change that state by the application of some external force. This form of statement applies to motions of translation. For motions of rotation, the law states that each body continues in a state of uniform motion of rotation about a fixed axis unless acted upon by some impressed force applied at some point not on the axis of rotation. Another relationship, known as Newton's second law of motion, states that the net or effective force acting on a body is proportional jointly to the mass and the acceleration produced by the force. A third relationship identified by Newton and known as Newton's third law states that for each force there is always an equal and opposite reaction.

An important property of matter, known as inertia, appears in Newton's first law of motion as that property of matter which causes it to remain at rest or in uniform motion unless affected by some outside force. There are two kinds of inertia because there are two general types of motion—translation and rotation. Bodies show opposition to being translated and also opposition to being rotated. The former opposition is called linear inertia; the latter is called rotary inertia. The linear inertia of a body is proportional to the mass of the body, but rotary inertia depends on the distribution of the mass about the axis of rotation. The relationship between inertia and mass can be stated in the form of an equation. Certain properties of moving bodies depend jointly on the mass and the velocity. This property is called the momentum of the body, and it is defined as the product of the mass and the velocity, which can be stated as:

$$\text{momentum} = \text{mass} \times \text{velocity} \tag{5.1}$$

Newton's second law can be expressed as:

$$\text{force} = (\text{a constant}) \times \text{mass} \times \text{acceleration} \tag{5.2}$$

More generally, Newton's second law states that the time rate of change of momentum is proportional to the impressed force.

Newton's third law can be expressed as:

$$\text{force} = \text{reactive force} \tag{5.3}$$

Other important concepts in physics include those of work and energy. Work is defined in terms of a relationship to force and distance: work = force × distance. Energy is defined as a body's capacity to perform work and consists of two types: potential and kinetic. Potential energy is that which a body has by virtue of its position or configuration. Kinetic energy is that which a body has by virtue of its motion. There are equations for each of these energy types that express their relationship to position and motion.

The relationships and equations described above apply equally well to all bodies—from molecules to planets. Since molecules are the smallest sizes of nonliving systems that retain their characteristics, they are the nonliving individuals treated in this book.

The quantitative science of heat was briefly mentioned in Chapter 1 and measures of temperature and heat energy were identified. Although heat is a form of energy and may be measured in the units in which mechanical energy is measured, it is convenient to use a unit which is based on the effects of heat in raising the temperature of a substance. In looking for a substance to use as a standard, it is natural to choose water because of the ease with which it is obtained. Since it always takes the same amount of energy to raise the temperature of one gram of water from 15° to 16°C, it is possible to define an arbitrary unit with which to measure other quantities of heat. The choice of the unit is a matter of convenience. In this respect, however, it does not differ from the unit of length or the unit of mass, which are also chosen arbitrarily. The unit of heat in the c.g.s. system of measure is the calorie. It is defined as the quantity of heat or energy necessary to raise the temperature of one gram of water from 15° to 16° on the centigrade scale. Because the heat required to raise the temperature of one gram of water 1°C is not the same at all temperatures, it is necessary to state the temperature at which the calorie is defined. This definition of heat quantifies the relationship between temperature and heat.

There is a relationship between temperature, heat energy, and the structure and organization of a substance. The amount of heat required to raise the temperature of one gram of another substance 1°C may be compared with the amount of heat required to raise the temperature of one gram of water from 15° to 16°C. Such a comparison gives a definition of the specific heat of the substance: the specific heat of a substance is equal to the number of calories required to raise the temperature of one gram of the substance 1°C.

Two other important relationships in heat science are known as the first and second laws of thermodynamics. The first law of thermodynamics is a special case of the law of conservation of energy. This first law of thermodynamics states that in the transformation of work into other forms of energy or in the transformation of one form of energy into other forms of energy, no energy is ever created or destroyed. That is, the energy before and after the transformation is always the same. The second law of thermodynamics states the conditions under which heat may be transferred from one body to another. It is, in effect, a statement of the fact that heat naturally flows from a place of higher to one of lower temperature but never in the reverse direction. The following is one form of statement of the law: It is impossible for any kind of a machine working in a cycle to transfer heat from a lower to a higher temperature unless external work is done on it.

The quantitative electrical science is shown in Chapter 1 as evolving from the fundamental measure of charge and the determinants resulting from the then extant quantitative sciences of mechanics and heat. Early in the evolution of the electrical science, it was discovered that there are two kinds of electricity: positive electricity and negative electricity. It was also discovered that there is a relationship between these two types of charges—charges which are alike repel each other, and charges which are unlike attract each other. This relationship was further developed into the law of electric force. If two point charges of electricity of opposite kinds are in the neighborhood of each other, they will exert attractive forces on each other. If they are of the same kind, they will exert repulsive forces on each other. The force which one exerts on the other is determined by the distance between the charges and the magnitude of the charges. This relationship was further specified in equation form: The force between the two point charges is inversely proportional to the square of the distance between the charges and directly proportional to the product of the charges. This law of force gave a method of defining the unit of electrostatic charge. Unit electrostatic charge is defined to be that charge which, when placed one centimeter from an equal charge of like kind in a vacuum, will repel it with a force of one dyne, sometimes called the statcoulomb.

5.3. Living Systems' Behavioral Relationships

The relationships among the determinants of living systems' behaviors were established following the procedures of the evolution of a quantitative science model. These determinants are behavioral energy, capacity to direct energy, behavioral information, and available energy. The relationships among these determinants are limited herein, to the readily observed behaviors of animals, because the model will eventually be applied to the human species.

Identification of the determinants of behavior in earlier chapters established the fact that there is a functional relationship between behavior and each of the determinants. That is, behavior is a function of a living system's capacity to direct energy, the energy available to the system, and the system's information. This functional relationship can be expressed as:

$$b = f(k, e_a, i) \qquad (5.4)$$

where b =behavioral energy, k = capacity to direct energy, e_a = available energy, and i = behavioral information.

The specific form of the relationship in Eq. 5.4 between a living system's behavior, its capacity to direct energy, its available energy, and its information was investigated first by considering limiting conditions, that is, can a living system exhibit behaviors when any of the fundamental parameters of behavior is missing? From the considerations in Chapter 3, it is evident that if a system does not have a capacity to direct energy, it cannot exhibit a behavior. From discussions of available energy in both Chapters 2 and 3, it is evident that if there is no available energy, there cannot be a behavior. From Chapter 4, it is evident that a living system cannot exhibit behavior unless there is an information input to cause a behavior. From these considerations, for a living system to exhibit behavior, it must simultaneously have all the following: a capacity to direct energy, available energy, and information. The functional form of an equation that meets these conditions is:

$$b = f(ke_a i). \qquad (5.5)$$

This relationship can be stated as follows:

The behavioral energy of a living system is a function of the product of the system's capacity to direct energy, the energy available to the system, and its behavioral information.

Analysis of Eq. 5.5 reveals that if any of the parameters k, e_a, or i is missing, there cannot be a behavior. This equation fits the observed facts as presented in Chapters 2, 3, and 4.

5.4. Quantitative Living Systems' Behavioral Relationships

Once the relationship among the determinants of living systems' behaviors is established, the next step in the evolution of the quantitative science model is to develop behavioral relationships that can be quantified. Because quantification of neural behavioral information has been developed at the motor unit level (Simms, 1991), the motor unit was used to establish the quantitative relationship for behavior. Indeed, the motor unit is the basic behaving element of the most obvious behaviors of animals. It is shown later that the behaviors of individual motor units can be combined into muscle behaviors and then to total animal behavior resulting from muscle contractions.

The contraction behavior of a motor unit, as characterized by the amount of energy converted by the unit during each contraction, is determined by three fundamental parameters. Each of these fundamental determinants—capacity to direct energy, available energy, and information—is treated herein for a motor unit and its relationship to behavior is specified in a form that can be quantified.

The following treatment is based, in part, on "The Fundamental Equations for the Behaviors of Living Systems" (Simms, 1990).

5.4.1. Capacity to Direct Energy

As shown in Chapter 4, a major characteristic of motor units is their capacity to convert chemical energy into work and heat energy each time the muscle fibers contract in response to a nerve impulse. The amount of energy converted during a contraction is a function of the number and types of muscle fibers in the motor unit. For example, a motor unit with only one muscle fiber will convert less energy during a contraction than a motor unit with a larger number of the same type fibers. Also, a motor unit composed of fibers associated with skeletal muscle will convert a different amount of energy than a motor unit with the same number of fibers of another fiber type, such as those associated with the smooth muscles of the intestines. That is, each specific type motor unit has its own unique capacity to convert chemical energy to work and heat energy and makes an energy conversion when "directed" to do so by a nerve impulse. Because the amount of energy converted by a specific motor unit is unique, it characterizes the behavior of the specific motor unit. The amount of energy in a motor unit's contraction behavior is a function of the particular motor unit's capacity to direct energy and can be expressed mathematically:

$$b_e = f(k_u) \tag{5.6}$$

where b_e = contraction behavior energy of a motor unit and k_u = a specific motor unit's capacity to direct energy.

The relationship between a motor unit's contraction energy and its capacity to direct energy was determined by analyzing the energy flow during contraction of the motor unit. Carlson *et al.* (1963) measured the total contraction energy (work + heat) of healthy non-fatigued muscles under conditions of maximum stimuli. Non-fatigued muscles have sufficient chemical energy available to each motor unit so that available chemical energy does not influence the relationship between a motor unit's contraction and its capacity to direct energy. The maximum stimuli condition ensures that all the motor units in the muscle are stimulated and therefore contract. Under these conditions, and using the conservation of energy principle, which states that energy is neither created nor destroyed, the contraction energy is equal to the motor unit's capacity to direct energy. That is, there is a direct relationship between contraction behavior energy and a motor unit's capacity to direct energy. Therefore, Eq. 5.6 becomes:

$$b_e = k_u \qquad (5.6)$$

Carlson *et al.* developed a methodology for measuring the total energy (work + heat) in a muscle contraction and, using this methodology, measured the total energy in a contraction of Rana pipiens paired sartorii. Since the amount of work performed by a muscle is a function of the amount of load on the muscle, Carlson *et al.* measured the total energy in a muscle contraction under a number of load conditions. Their measurements were performed on the paired sartorii muscle using a number of Rana pipiens muscles of differing sizes. To negate the differences in the muscle sizes, they normalized the measurement data to obtain the total energy per gram of muscle per contraction as a function of the load on the muscle. A better comparison of muscle size is the cross-sectional area of the muscle as opposed to weight. However, it is much easier to measure the weight of muscle than it is to measure its cross-sectional area. Carlson and Wilkie (1974) report that the total energy per gram is essentially the same as the total energy per square centimeter of muscle cross-sectional area. They normalized the load P on the muscle by expressing it as a fraction of the peak isometric (constant length) contraction tension p_{ot}. A curve of the total amount of energy in a single contraction of Rana pipiens paired sartorii muscle is shown in Figure 4.1 in Chapter 4.

The data from this curve were used to construct a figure that more clearly depicts total energy per contraction as a function of load. The total energy per contraction per motor unit is shown in Figure 5.1 for the Rana pipiens satorius motor unit. This total energy is the motor unit's capacity to direct energy. It can be seen from this figure that the capacity to direct energy is a function of the load. Figure 5.1 shows a Rana pipiens sartorius motor unit's capacity to direct energy, as a function of the load on the motor unit.

The procedure of Carlson and Wilkie—along with the author's method for converting from muscle mass to motor units—can be used to measure the capacity to direct energy of other Rana pipiens motor units. These procedures can also be used to measure the capacities to direct energy of the motor units of other animal species.

5.4.2. Available Energy

When there is ample chemical energy, a motor unit will convert energy at its capacity to direct energy. Due to the conservation of energy principle, as chemical energy in the motor unit is depleted, the energy converted by the motor unit to work and heat is reduced. This phenomenon is observed when a muscle fatigues, which is a manifestation of chemical energy depletion in the muscle. This phenomenon is expressed as:

$$b_e = f(e_a): \qquad 0 < e_a < e_m \qquad\qquad (5.8)$$

where e_a = available energy and e_m = the maximum chemical energy that can be converted to work and heat energies during each contraction, and is unique to a specific motor unit. This equation states that behavioral energy is a function of available energy and is valid for available energy values between zero and the available energy used for the maximum contraction of the motor unit. This maximum energy is a function of the structure and organization of a specific motor unit. The form of this functional equation was obtained from energy considerations. Due to the conservation of energy principle, a reduction in available energy results in a proportional reduction in behavioral energy, with all other parameters remaining constant. Under these conditions, Eq. 5.8 becomes:

$$b_e \propto e_a: \qquad 0 < e_a < e_m \qquad\qquad (5.9)$$

This equation states that behavioral energy is proportional to available energy for the range of available energies from zero to a specific maximum. This maximum available energy is equal to the capacity to direct energy of the motor unit under consideration.

5.4.3. Information

A motor unit converts chemical energy to work and heat energies only when caused to do so by a nerve impulse. In living systems with central nervous systems, the nerve impulses which cause motor unit contractions can be the result of information processing in the central nervous system. At the motor unit, there is either no message or there is a message for the muscle fibers to contract. In essence, processed information from the central nervous system "directs" the motor unit to contract and thereby convert chemical energy into work and heat. The processed information from the central nervous system is neural information.

A motor unit's contraction behavior exists only when there is a neural information input. Therefore, the contraction behavior is a direct function of this behavioral information. That is:

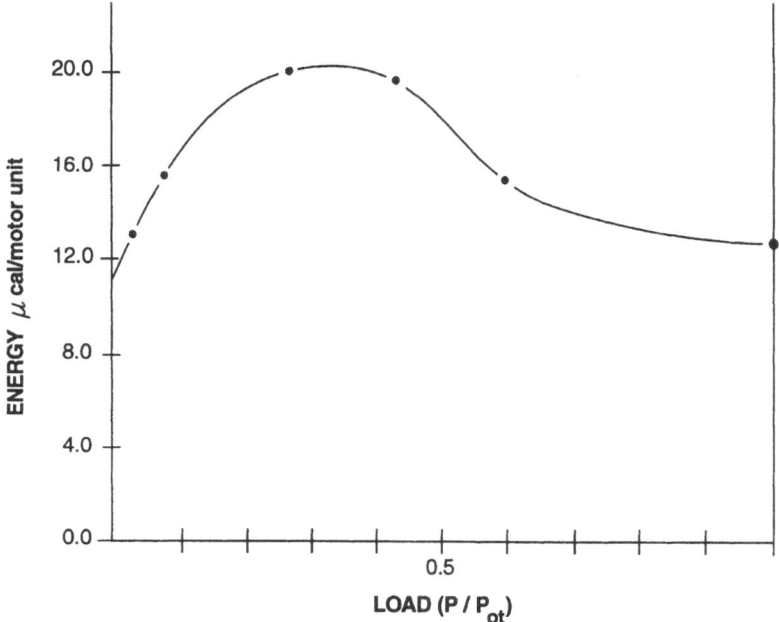

Figure 5.1. Capacity to direct energy vs. load.

$$b_e = f(i_u) \tag{5.10}$$

where i_u = behavioral information input to a motor unit.

The information input that causes a motor unit contraction is in quantum form. That is, there is either no information input or there is an information input which activates the motor unit's muscle fibers, and the input is in the form of an impulse. When there is a quantum of behavioral information, that is, an action potential, the motor unit exhibits contraction behavior. Therefore, the motor unit's behavior is a direct function of quantum information inputs and can be expressed as:

$$b_e \propto i_u: \qquad i_u = 0 \text{ or } 1 \text{ quantum} \tag{5.11}$$

5.4.4. Fundamental Equations of Motor Unit Behavior

Equations 5.6, 5.8, and 5.10 can be combined to express a motor unit's contraction behavior as a function of its capacity to direct energy, the available energy, and behavioral information. Motor unit contraction only occurs when the motor unit has a capacity to direct energy, when there is chemical energy available for conversion to work and heat, and when there is a quantum behavioral information input. That is:

$$b_e = f(k_u, e_a, i) \qquad (5.12)$$

Equations 5.7, 5.9, and 5.11 state that a motor unit's contraction behavior is directly related to k_e, e_a, and i. Therefore, Eq. 5.12 becomes:

$$b_e = f(k_u e_a i)$$

The general form of this direct relationship equation is:

$$b_e = A k_u e_a i + C : \qquad 0 < e_a < e_m \qquad (5.13)$$

where A and C are constants. The exact form of this equation is determined by considering muscle contraction characteristics and the experimental data of Carlson *et al.*. Since contraction behavior (b_e) does not exist when any of the parameters k_u, e_a, and i are zero, the constant C must be zero. For any given load on the motor unit, the unit's capacity to direct energy is constant and the conversion of a given amount of available energy into contraction behavior energy is determined only by this capacity to direct energy. Therefore, the constant A in Eq. 5.13 is equal to one and the equation becomes:

$$b_e = k_u e_a i : \qquad 0 < e_a < e_m \qquad (5.14)$$

A motor unit has specific characteristics which allow the further specification of Eq. 5.14. A motor unit has an "all or none" behavioral response to a quantum of information input (Lucas, 1917), provided there is sufficient chemical energy available. Under these normal conditions, the energy used in a motor unit contraction is at the motor unit's capacity to direct energy. Equation 5.14 becomes:

$$b_e = k_u i : \qquad \text{given that } e_a > k_u \qquad (5.15)$$

This equation states that the behavioral energy in a motor unit contraction is equal to the motor unit's capacity to direct energy in a single contraction, given that the chemical energy available to the motor unit is equal to or greater than the motor unit's capacity to direct energy.

The behavior of a Rana pipiens sartorius motor unit can be used to illustrate the application of Eq. 4.15. Using the Rana pipiens sartorius motor unit data given in Chapter 4, the motor unit has a capacity to direct 18.5 microcalories of energy per quantum information input when it is working against a load that is

0.45 of its maximum load and there is adequate chemical energy available, that is, the muscle is not fatigued. Under these conditions, the parameters in Eq. 5.15 are:

$$k_u = 18.5 \text{ microcalories per quantum of behavioral information}$$

$$i = 1 \text{ quantum of behavioral information}$$

The behavioral energy, from Eq. 5.15 is:

$$b_e = 18.5 \times 1 = 18.5 \text{ microcalories}$$

The measurement unit check of Eq. 5.15 shown below demonstrates that the units are consistent:

$$b_e = k_u i$$

$$\text{energy} = (\text{energy/quantum}) \times \text{quantum} = \text{energy}$$

This example is for a capacity to direct energy that is associated with a particular load on the motor unit. As seen in Figure 5.1, a motor unit's capacity to direct energy is a function of the load on the motor unit. A polynomial equation for a Rana pipiens sartorius motor unit's capacity to direct energy as a function of load conditions can be derived based on the data presented in this figure. Similar polynomial equations can be developed for other motor units' capacities to direct energy as a function of load, provided experimental data are available.

Equation 5.15 was modified to account for the dependence of a motor unit's capacity to direct energy on its load. For a particular load, this equation becomes:

$$b_{el} = k_{ul} i : \quad \text{given that } e_a > k_{ul} \tag{5.16}$$

where b_{el} = a motor unit's behavioral energy for a specific load l on the motor unit, and k_{ul} = a motor unit's capacity to direct energy for a specific load l on the motor unit.

This equation is valid for any load from zero up to the peak isometric tension of the motor unit.

The total energy in a Rana pipiens motor unit contraction consists of two parts: heat and work. Equation 5.15 can be expanded to explicitly show these two behaviors as follows:

$$b_w + b_h = (k_{uw} + k_{uh}) \times i : \quad \text{given that } e_a > k_u \tag{5.17}$$

where: b_w = behavioral work energy for a given muscle load
b_h = behavioral heat energy for a given muscle load
k_{uw} = the motor unit's capacity to direct work energy for a given load
k_{uh} = the motor unit's capacity to direct heat energy for a given load

When a motor unit fatigues, that is, when chemical energy is lower than that necessary for normal behavior, Eq. 5.15 is no longer valid. Equation 5.15 can be modified to account for available energy depletion by considering two limiting conditions. The first is when available energy and behavioral energy are zero. The second is when available energy is equal to the motor unit's capacity to direct energy and the behavioral energy is normal. An equation which meets these conditions is:

$$b_e = k_u \times (e_a/k_u) \times i: \quad 0 < e_a < k_u \tag{5.18}$$

This equation is valid for available energies from zero up to a level where there is sufficient energy for a full contraction of the muscle fibers of the motor unit.

Equations 5.16 and 5.18 are the fundamental equations for motor unit behaviors. These equations provide the tools for quantifying the behaviors of the basic behaving element of animals with neural systems—the motor unit.

Equations 5.16 and 5.18 give the behavioral energy for each quantum of behavioral information. The energy for a number of information quanta is obtained by adding the behavioral energies resulting from each information input. There are limits to the quanta of information that can be input to a motor unit in a given period of time, that is, a limit on the information rate. The information rate limit is due to the recovery time of the motor unit's motoneuron after an electrochemical pulse, which transports the quantum of information, has traveled through the motoneuron.

5.5. Specific Capacity to Direct Energy

The quantitative science evolution model presented early in this chapter describes the concept of specific heats. This concept is based on substances having a unique thermal capacity which is a function of their structure and organization. The specific heat of a substance is the ratio of its thermal capacity to the thermal capacity of water. The thermal capacity of a substance is the number of calories needed to raise one gram of the substance by 1°C. The specific heat of substances has been useful in the heat sciences to describe this fundamental thermal capacity of substances.

The model for specific heat suggested that a similar concept for comparing the capacity to direct energy of motor units to the capacity to direct energy of the reference system—the Rana pipiens sartorius motor unit—would be useful. Such a model can provide a comparison of the capacity to direct energy characteristics among different motor units, just like specific heats of substances provide a means for comparing heat capacities of substances. Because the model had served so well in the development of the fundamental equations for living systems' behaviors, a concept for the specific capacity to direct energy for motor units was developed.

The definition of specific heat as the ratio of the thermal capacity of a substance to the thermal capacity of water provides a good model for defining the specific capacity to direct energy of motor units. A motor unit's specific capacity to direct energy is the ratio of its capacity to direct energy to the capacity to direct energy of a motor unit of Rana pipiens sartorius muscle. Because the reference Rana pipiens motor unit has a capacity to direct 18.5 microcalories of energy, the specific capacity to direct energy of other motor units is the ratio of their capacity to direct energy to the 18.5 microcalories directed by the Rana pipiens reference motor unit. The capacity to direct energy of other motor units to be used in the calculation of specific capacity to direct energy must be measured under the same conditions that were used to measure the 18.5 microcalories directed by the reference motor unit. These conditions include temperature of the motor unit, load on the motor unit as a fraction of its peak isometric load, sufficient available energy to the motor unit, an information input that will cause contraction of the motor unit, and the directed energy measured in microcalories.

The specific capacity to direct energy of motor units provides a method for comparing this fundamental behavior characteristic of motor units. It should also provide a means for easier computations of motor units' behaviors. If the specific capacities to direct energy prove useful, there will likely be tables of specific capacities to direct energy developed for the various motor units of animals, just as there are now tables of specific heats of substances published in handbooks.

5.6. Multiple Motor Units

Single motor units are sometimes found in very simple animals and sometimes in more complex animals. Usually, they are found combined with other motor units to form a muscle. For example, approximately 16 motor units combine to form the Rana pipiens sartorius muscle. The contraction behavior of a muscle is determined by the number and types of motor units in the muscle, how the motor units are structured to form the muscle, and the combinations of neural information inputs to the muscle's motor units. As an illustration, the Rana pipiens sartorius muscle is of the striated type associated with skeletal muscle. It consists of long muscle fibers and approximately 16 motor units. Each motor unit can be activated separately to give a graduated muscle contraction.

The behaviors of a muscle consisting of multiple motor units can be quantified using the characteristics of the muscle and Eq. 5.16. A muscle is characterized by (1) the number of its motor units, (2) the type of motor units (e.g., length, cross-sectional area, mass, and tissue structure), (3) the way the motor units are combined to form the muscle, (4) the motor units' capacities to direct energy as a function of load, and (5) the ways information can be transmitted to the muscle. Muscle behavior is a function of these characteristics and the specific information transmitted to the muscle. This information input to the motor units, as a function of time, is required to obtain the behavior of the muscle over some time period. The resulting time-dependent motor unit behaviors caused by information inputs can be determined and then combined to obtain the behavior of the muscle.

As motor units combine to form individual muscles, so do muscles combine into muscle groups which in turn combine into networks of muscle groups. Thus, they form, for example, complete skeletal muscle systems of an animal. The smooth muscle systems, such as those associated with the circulation of body fluids, are also formed from motor units and muscle groups. Given the ways motor units are combined into muscles, the equations for the behaviors of motor units can be used to quantify the total behavior of muscle groups and, indeed, the behaviors of the individual animal. This will be discussed further in following chapters.

5.7. Summary

Quantitative description, analysis, and prediction of muscle behaviors require an understanding of the characteristics and processes of the motor unit—the basic muscle behavior system. The fundamental characteristics and processes of a motor unit are shown schematically in Figure 5.2.

The amount of chemical energy converted to work and heat energy each time there is a behavioral information input is a function of the particular motor unit's capacity to direct energy (k_u).

The contraction behavior of a motor unit, as described by Eq. 5.16 and 5.18, is applicable to a single motor unit and to a motor unit in multi-motor unit muscles. The contraction behavior of a motor unit over some period of time is a function of the number and spacing of the behavioral information input to the motor unit over this same period of time. The behavioral information input can vary from zero to the maximum information rate that the system can accept.

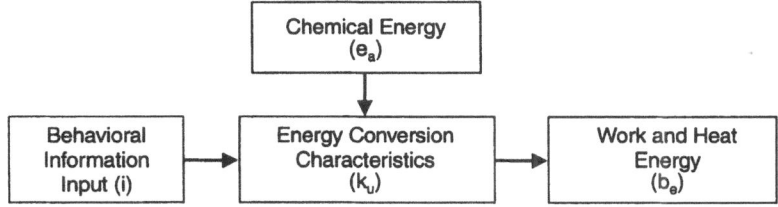

Figure 5.2. Motor unit characteristics and processes.

The behavior of muscles consisting of two or more motor units acting in a coordinated way is a function of information processing prior to the behavioral information being transmitted to the various motor units. In lower-level animals with few muscles and motor units, the behaviors and information-processing functions are relatively easy to quantify and analyze. However, in higher-level animals with many muscles and motor units, the behaviors and information processing may be very complex.

The methodologies and equations developed herein can be used to measure, analyze, and predict the muscle behaviors of living systems. They also provide the theoretical foundation for understanding the observed behaviors of living systems with motor units.

CHAPTER 6

Living Systems Science Evolution

6.1. Introduction

The evolution of a quantitative science includes not only the development of the basic principles of the science but also the testing, verification, and application of these principles. Because the basic principles of a quantitative science emerge from preceding qualitative science, these principles are partially verified through the development and application of this preceding science. This chapter illustrates how some of the extant quantitative sciences evolved from their preceding qualitative phase and uses the evolution of these sciences as a model for treating the evolution of a quantitative living systems science. Because the extant quantitative sciences have passed through the basic principles stage of evolution, their evolution can be used as a model for evolving a living systems science from the basic principles which were developed in the preceding chapters. A model based on extant quantitative sciences is developed in this chapter and used in identifying the evolution of living systems science from its earliest beginnings to its current state. This model also provides guidance for the verification and application of the basic principles of quantitative living systems science.

Before proceeding to the testing, verification, and application phase in the evolution of a quantitative living systems science, a clear statement of the basic principles to be verified and applied is given. The basic principles of quantitative living systems science developed in the preceding chapters are concisely stated as follows:

- The behaviors of living systems can be observed via the energy utilized in these behaviors.
- The energies utilized in living systems' behaviors can be measured or calculated—therefore, living systems' behaviors can be quantified.
- A living system has its own unique behavioral characteristic, its capacity to direct energy.
- A living system's capacity to direct energy is a function of its structure and organization.

- A living system's behavior is a direct function of the energy available to the system.
- A living system's behavior is a direct function of behavioral information.
- Behavioral information is the ability to cause work (the utilization of energy) and can be measured or calculated by the work it causes.
- A living individual's behavioral information is in the form of genetic and chemical information. Animals also have neural information.
- A living system's behavior is a direct function of its capacity to direct energy, available energy, and behavioral information.

After completing the fundamental theory development phase (development of fundamental measures, basic principles, and equations), the research was shifted to the next phase in the evolution of a quantitative living systems science. This phase consisted of testing the basic principles using extant data. The purpose of testing is to verify the principles and to establish a high degree of confidence in these principles.

The working hypothesis for the next phase of a quantitative living systems science evolution is that the processes and procedures used in the evolution of the quantitative nonliving systems is applicable. Although the model of the evolution of the quantitative sciences, as described in Chapter 1, served well for developing the fundamental principles of a quantitative living systems science, a broader and more detailed model is required for the next phase of a quantitative living systems science.

A brief overview of this new model (Simms, 1993) is given here because it provides the structure for the development of a quantitative living systems science. The development of this model is given later in this chapter. A major feature of the model is that the evolution of a quantitative science proceeds from ideas, to concepts, to hypotheses, to theories, and to laws as shown in Figure 6.1. Another major feature is that the confidence levels in a science increase as the science evolves through various stages of ideas, concepts, hypotheses, theories, and laws. Figure 6.1 illustrates this increase in confidence levels as a science proceeds from the idea stage to the law stage. In the actual evolution of a science, the demarcation between stages is often blurred. Curves of confidence levels versus time or effort would be different from that shown in the figure, because the time span or the effort associated with the stages may vary for the various sciences.

A definition of the word *idea* includes a formulated thought or opinion as well as whatever is known or imagined regarding an object. In the idea stage, the confidence level in the formulated thoughts or opinions can be close to zero. Confidence levels are used herein in the mathematical sense and can range from a complete lack of confidence (zero) to complete confidence (one).

As ideas evolve into concepts, confidence levels in an evolving science are increased. A *concept* is defined as something conceived in the mind and an abstract idea generalized from particular instances. In this stage, one or more con-

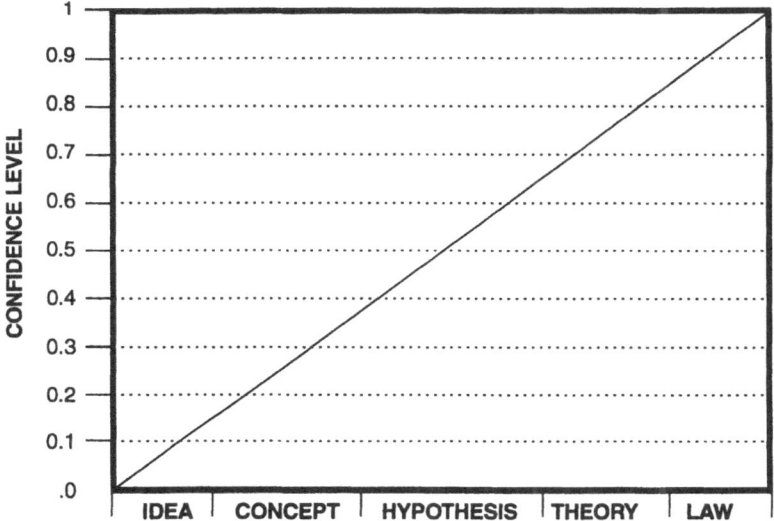

Figure 6.1. Confidence in stages of quantitative science evolution.

cepts may be synthesized which embody and expand the formulated thoughts or opinions on the idea. These concepts are constructed so they can be verified to some degree using existing facts and analytical and technical tools. The degree of confidence in these concepts is a function of the extant facts, the validity of the analytical and technical tools, and the applicability of these tools.

The next stage in increasing confidence levels in ideas and concepts is to test the ideas under various hypothetical conditions. These conditions are the result of the hypothesis of factors which affect the conceptual systems. A hypothesis is a tentative assumption made in order to draw out and test its logical or empirical consequences, and an assumption or concession made for the sake of argument. Testing a conceptual system under various hypotheses increases the confidence level in this system, provided the hypothesis successfully passed the tests.

Definitions of the word *theory* include the analysis of a set of facts in their relation to one another, the general or abstract principles of a body of fact, a plausible or scientifically acceptable general principle or body of principles offered to explain phenomena, and a body of theorems presenting a concise systematic view of a subject. These definitions of theory imply a much higher confidence level in this stage of the evolving science than in the previous stages. The last stage of a science is the evolution of laws applicable to the subjects of the science. Definitions of the word *law* include a statement of an order or relation of phenomena that, so far as is known, is invariable under the given conditions, a relation proved or assumed to hold between mathematical expressions, and the observed regularity of nature. In this stage, the principles of the science have the highest confidence levels.

Following the section in this chapter on the development of the evolution model is a section describing the current state of a quantitative living systems science in relation to this evolution model. This latter section identifies where the science is, the next stage in the development, and the scope of the work necessary to continue the evolution of the science.

6.2. Expanded Evolution Model

The major purpose of the quantitative science evolution model developed in Chapter I is to relate the quantitative sciences to fundamental measurements and to show the interrelationships among the quantitative sciences. The major purposes for expanding this model are to provide a broader view of the evolution of the quantitative nonliving systems sciences and to develop a model that can be used for further development of a quantitative living systems science.

The broader or more comprehensive model (Simms, 1993) for the evolution of the quantitative sciences is based on the processes and procedures associated with the evolution of science and technology. In developing this model, it was postulated that the fundamental processes and procedures used in the evolution of science are similar over long periods of time. Of course, these processes and procedures are always being refined as new tools and instruments are developed. For example, the concept of confidence level has changed from one based on casual observation to the precise mathematical methods available today for calculating confidence levels. The basic process for the development of a quantitative science, or the principles thereof, starts with an idea and ends with verified truths. The process consists of the following major steps: (1) idea (the generation of an idea, or ideas, about a subject); (2) concept (elaboration of the idea, or ideas, into concepts which have more specificity and structure); (3) hypothesis (establishment of hypotheses which allow testing and a degree of validation of the concepts); (4) theory (developing theory which establishes general principles); and (5) law (verification of relationships and rules).

The postulated model was tested and validated using a historical perspective of the quantitative sciences and observing the common elements associated with the evolution of these sciences. The quantitative sciences of geometry, astronomy, and chemistry were used as the basis for developing the model. Evolution of the quantitative science of mechanics based on the postulated model has been reported previously (Simms, 1993).

Development of the model starts with the broadest definitions of geometry, astronomy, and chemistry. Geometry is the science of the properties of space; astronomy explores the properties and behaviors of everything beyond the earth;

and chemistry is the study of the composition and structure of substances and the changes they undergo. These broad definitions allow consideration of these evolving sciences back to prehistoric times.

Pre-recorded history artifacts demonstrate the importance to human endeavors of the properties of space, the properties and behaviors of things beyond the earth, and the composition and behavior of substances. The identification, location, and protection of hunting and gathering territories are crude and inexact beginnings of surveying, a precursor to the beginnings of geometry. The properties and behaviors of things beyond the earth, such as the energy of the sun, were important to prehistoric humankind, just as they are important today. Artifacts from the Stone Age indicate the importance of the composition of materials to early humankind.

An observation based on artifacts of the distant past is that the quantitative sciences have been evolving over a very long period of time and that the subjects of these sciences are, and have been, important to human endeavors. Another observation is that these so-called hard sciences (quantitative sciences) all evolved from soft (i.e., non-quantitative) sciences.

The model is based on recorded data, that is, on the recorded history of humankind and on artifacts. The period considered starts about the time of the emergence of civilization. These data are incomplete and biased on the side of those ideas which were successful in the sense that they survived, were useful, and therefore recorded. However, these data are all that is available.

Development of the model is described below by treating geometry first, then astronomy, and then chemistry. The common elements, as they affect the model, are identified following these individual treatments.

6.2.1. Geometry

The processes by which early empirical geometry evolved from prehistoric concepts of space probably can never be known with a high degree of confidence. However, some reasonable estimates of the processes can be made based on the outcomes of these processes. The definition of the word geometry provides a starting place. The Greeks coined this word from the Greek words for earth and measure. Prerequisites for geometry are that certain figures can represent space and that units of measure must exist. Some human or humans had to have the idea of combining geometric shapes and measures. Early Babylonian records disclose the forerunners of triangles, quadrangles, and parallel lines. Drawings of Babylonian carriages indicate the division of a circumference into four and six equal parts. For angular measurement the circumference was divided into 360 equal parts or degrees. One tablet reveals the approximate computation of the diagonal of a rectangle.

It is generally believed that empirical geometry started in ancient Egypt. This early empirical geometry consisted merely of a number of crude rules for the measurement of various simple geometric figures. The earliest empirical geometry was associated with shapes in a plane. However, empirical geometry also treated shapes of three dimensions. For example, an early papyrus from ancient Egypt gives a calculation of the exact volume of the frustum of a rectangular pyramid.

From the perspective of the development of an evolution model, the prerequisites for empirical geometry were (1) the idea for combining geometric figures and measurement, (2) the concept that geometric figures could represent land surfaces, (3) the testing of the concept, (4) development of beliefs that the concept could be generalized to work in a number of circumstances, and (5) the generation of crude rules (inexact laws) which could be used to survey land with a reasonable degree of accuracy and confidence. This high confidence can be implied from the long usage of these inexact laws. These crude rules were used prior to their being described by the Egyptian priest Ahmes (Newman, 1956) considerably more than a thousand years before the beginning of the Christian era. They were used until after the emergence of demonstrative geometry initiated by the Greek Thales of Miletus who lived about 640–546 B.C.

Demonstrative geometry depends on the idea and observation that the propositions of geometry are logically interrelated, that is, that if certain propositions are granted, certain others can be proved as logical consequences of those assumed. From the evolution point of view and from hindsight, this geometry must have emerged from the idea that the logic concepts of the Greeks could be combined with empirical geometry to produce a more logical geometry. Thales' work, as can be inferred from Proclus, consisted of a number of propositions whose proofs were deductive as opposed to inductive, as was probably the case with the Egyptian geometry. Thales introduced the idea of establishing by exact reasoning the relations between the different parts of a figure, so that some of them could be found by means of others in a rigorous manner. This was a phenomenon quite new in the world. The central idea of Euclid's work is that the propositions of geometry could be put into a systematic, logical structure. It is one of the marvels in the history of mathematics that the *Elements*, written by Euclid in the fourth century B.C., was used as a textbook in geometry for over 2,000 years. The main ideas of demonstrative geometry are that the proposition of geometry could be logically interrelated and could be proved by deductive reasoning. For these ideas to become accepted as "laws," they had to be synthesized into concepts, and the concepts had to be tested to the extent that they were believed to have an extremely high confidence of being correct.

Demonstrative geometry was followed by the invention of analytical geometry in the early 1600s. The two main ideas involved in analytical geometry are the location of points in a figure by the use of coordinates and the algebraic representation of a curve or surface by an equation involving two or three variables. The concept of combining the extant geometry, algebraic representation, and coordinate representation was a new synthesis of geometry. The chief credit for the invention of analytical geometry belongs to René Descartes. He successfully tested his idea and concept by solving the "Problem of Pappus." This problem offered an excellent example of the power of the analytical method. During the eighteenth century, the ideas of calculus were introduced to analytical geometry, which greatly increased its power. The introduction of calculus into analytical geometry resulted in the discipline known as differential geometry.

Other geometries have evolved which are called non-Euclidean geometries. They are based on assumptions which differ from those of Euclid. The ideas resulting in the non-Euclidean geometries are very interesting and important but are not treated here because the previous geometries provide sufficient data to evaluate the evolution model.

Analysis of the long evolution of geometry, from the first artifacts such as drawings and solid carvings in ancient caves to the four-dimensional space–time manifold of Einstein and Minkowski in this century, reveals one consistent idea: Space, and the objects in space, can be represented by similar shapes of reduced dimensions. The ancient drawings on the walls of caves consist of approximate representations of the animals of that time. The mathematical geometry of Einstein is a representation of the space of the universe. As geometry evolved, these representations became more exact and useful in their description of real space and the objects therein. For example, the idea of ancient empirical geometry was the assignment of metrics to shapes, which allowed more scaling between the real world and the representations. The Egyptians introduced the idea that the geometric representations could be scaled up to recreate (survey) real land boundaries from these representations. The Greeks introduced the idea of logical relationships among the geometric representations, which made the construction of these representations more efficient and closer to the real world. Thus there was a high degree of confidence that the representations were precise. The ideas resulting in the invention of analytical geometry provided means for even better representations of the real world and a resulting increase in the confidence of these representations. The ideas of non-Euclidian geometry provided a means for better representations of the space of the universe. Confidence levels in the non-Euclidian representations of the universe have increased in the twentieth century.

Although the historical record of the processes and procedures used in the evolution of geometry is not complete, it seems clear that they are essentially the same as those in use today.

6.2.2. Astronomy

Astronomy evolved due to the practical needs of man. A practical understanding of the elements of astronomy is indispensable to the conduct of human life. It is widely used among tribal peoples whose existence depends upon the dictates of nature. Having no clocks, they depend instead on the characteristics of the sky. The stars serve them for almanacs. They hunt and fish, sow and reap by the recurrent order of celestial appearances. In China, Egypt, and Babylonia considerable skill was attained in the arts of celestial observation. The data from these observations were used to deduce empirical rules which increased the range of predictions and the accuracy of these predictions. But no genuine science of astronomy was founded until the Greeks combined the idea of logic with experience to develop theory.

Pythagoras of Samos learned during his travels in Egypt and the East, in the fifth century B.C., to identify the morning and evening stars, and to regard the earth as a sphere freely poised in space. The tenet of its axial movement was held by many of his followers—in an obscure form by Philoaus of Crotona after the middle of the fifth century B.C., and more explicitly by Ecphantus and Hicetas of Syracuse (fourth century B.C.), and by Heraclides of Pontus. Heraclides, who became a disciple of Plato in 360 B.C., taught that the sun, while circulating round the earth, was the center of revolution to Venus and Mercury. An actual heliocentric system, developed by Aristarchus of Samos in the second century B.C., was described by Archimedes in his *Arenarius*, only to be set aside with disapproval. At this stage in the evolution of astronomy, there were many conflicting ideas about the world outside the earth. Some additional ideas were that the earth was flat and that the earth was the center of the whole universe. The confidence level in these various ideas was low as demonstrated by their number and the lack of hard data to verify any of them. Like the heliocentric system concept, each of the ideas had its own concept of a planetary system—with equally low confidence levels.

The first mathematical theory of celestial appearances was devised by Eudoxus of Cnidus (408–355 B.C.). His idea was to combine uniform circular movements to represent observed appearances. The sun and moon and the five planets were accommodated each with a set of variously revolving spheres. The Eudoxian or "homocentric" system, was modified by Apollonius of Perga in the second century B.C. into the hypothesis of deferents and epicycles. This modified homocentric system was the embodiment of Greek ideas in astronomy for 1,800 years. Eudoxus also wrote two works descriptive of the heavens, the *Enoptron* and *Phaenomena*, which provided all the leading features of modern stellar nomenclature.

Greek astronomy culminated in the school of Alexandria where Aristyllus and Timocharis (ca. 320–260 B.C.) constructed the first catalogue giving star positions as measured from a reference point in the sky. This fundamental advance rendered inevitable the detection of precessional effects and firmly established the idea of quantification, or measurement, into astronomy.

Among the astronomers of antiquity, two great men stand out with unchallenged pre-eminence. Hipparchus and Ptolemy considered the same large organic designs, they worked on similar methods, and their results fit together so accurately that between them they remade celestial science. Hipparchus fixed the chief data of astronomy: the length of the tropical and sidereal years, length of the various months, and length of the synodic periods of the five planets. He also determined the obliquity of the ecliptic and of the moon's path, the place of the sun's apogee, the eccentricity of its orbit, and the moon's horizontal parallax, all with approximate accuracy. His supreme merit was in the establishment of astronomy on a sound geometrical and quantitative basis. His acquaintance with trigonometry, a branch of science initiated by him, together with his invention of the planisphere, enabled him to solve a number of elementary problems. These concepts of astronomy increased the degree of confidence greatly over previous concepts.

The choice made by Hipparchus of the geocentric theory of the universe decided the future of Greek astronomy. His catalogue of 1,080 stars, divided into six classes of brightness is one of the finest monuments of ancient astronomy. It is substantially embodied in Ptolemy's *Almagest*, which was the consummation of Greek astronomy.

The next major evolution of astronomy is attributable to Copernicus. He laid the groundwork of his heliocentric theory between 1506 and 1512 and brought it to completion in *De Revolutionibus Orbirm Coelestium* (1543). The task of remaking astronomy on an inverted design was, in this treatise, virtually accomplished. Its idea and reasoning were solidly founded on the principle of the relativity of motion. The continuous shifting of the standpoint was in large measure substituted for the displacements of the objects viewed, which thus acquired a regularity and consistency heretofore lacking. In Copernicus' system, the sphere of the fixed stars no longer revolved diurnally, the earth rotating instead on an axis directed toward the celestial pole. The sun, too, remained stationary while the planets, including the earth, circulated round the sun. By this means, the planetary "retrogradations" were explained as simple perspective effects due to the combination of the earth's revolutions with those of her sister orbs. The retention, however, by Copernicus of the antique postulate of uniform circular motion degraded the perfection of his plan since it involved a partial survival of the epicyclical machinery. Also, it was not feasible to place the sun at the true center of any of the planetary orbits. The reformed scheme was by no means perfect. Its simplicity was only comparative; many outstanding anomalies compromised its working. Moreover, a mobile earth outraged deep-rooted prepossessions. Under these circumstances, it is not surprising that the heliocentric theory won its way slowly to acceptance as the confidence level in this theory increased.

The *Tabulae Prutenicae*, calculated on Copernican principles by Erasmus Reinhold (1511–1553), appeared in 1551. Although the tables represented celestial movements far better than the Alfonsine Tables, large discrepancies were still apparent, and the desirability of testing the Copernican hypothesis upon which they were based by more refined observations prompted a change of methods. Tycho Brahe perfected the art of celestial observation with the instruments invented prior to the telescope. With these instruments and his great skills in their use, he obtained data of an unprecedented degree of accuracy for that period of time. Tycho never designed his observations to be an end in themselves. He thought of them as a means toward the end of ascertaining the true form of the universe.

Johann Kepler used the wealth of materials amassed by Tycho Brahe to test a number of hypotheses he had conceived. Kepler's work resulted in a series of unique discoveries. His computations persuaded him that the planets traveled in an ellipse, one focus of which was occupied by the sun. He found that a planet's velocity was uniform with respect to no single point within the orbit, but that the areas described, in equal times, by a line drawn from the sun to the planets were strictly equal. These findings were published in 1609 in *De Motibus Stellae Martis*. The announcement of the third of "Kepler's Laws" was made ten years later, in *De Harmonice Mundi*. It states that the squares of the periods of circulation round the sun of the several planets are in the same ratio as the cubes of their mean distance. The outcome of his discoveries was not only the perfection of the geometrical plan of the solar system, but the enhancement of the predicting power of astronomy. The Rudolphine Tables, computed by him from elliptical elements, retained authority for a century and in principle have never been superseded. The confidence level of Kepler's theories were increased so much by his work that the theories gained the status of laws.

Galileo Galilei was a contemporary of Kepler whose contributions to astronomy were different from Kepler's. They were easily intelligible to the general public; in a sense, they were obvious, since they could be verified by every possessor of telescopic instruments. Through his work, Galileo ranks as the true founder of descriptive astronomy. However, the accumulation of facts does not in itself constitute science. The foundations of a mechanical theory of the heavens were laid by Kepler's discoveries, and by Galileo's dynamical demonstrations; construction of the theory was facilitated by the development of mathematical methods, such as logarithms, analytical geometry, and calculus.

Jeremiah Horrocks had some intuition, prior to 1639, that the motion of the moon was controlled by the earth's gravity and disturbed by the action of the sun. Ismael Bouillaud (1605–1694) stated in 1645 that planetary circulation was under the sway of a sun-force decreasing as the inverse square of the distance. The inevitability of this same "duplicate ratio" was separately perceived by Robert Hooke, Edmund Halley, and Sir Christopher Wren before Newton's discovery had yet been made public. But Newton was the only man of his generation who both recognized the law, and had the power to demonstrate its validity. His achievement was twofold: first, the identification, by strict numerical comparisons, of terrestrial gravity with the mutual attraction of the heavenly bodies, and second, its mechanical consequences throughout the solar system. Gravitation was thus shown to be the sole influence governing the movements of planets and satellites. The figure of the rotating earth was successfully explained by its action on the minute particles of matter; tides and the precession of the equinoxes proved amenable to reasoning based on the same principle; and gravity satisfactorily accounted as well for some of the chief lunar and planetary inequalities.

Since Newton's time, there have been and continue to be major advances in astronomy. These advances resulted from better observation instruments, such as improved telescopes and spectroscopy instruments. Other advances have resulted from the application of atomic physics and the quantum theory to the conditions in stars and nebulae.

The history of the evolution of astronomy provides an excellent illustration of the processes by which a quantitative science evolves. The numerous and conflicting ideas relating to the nature of the heavenly bodies and their behaviors are indicators of the low confidence levels associated with these ideas. These competing ideas were formulated into conceptual systems. Hypotheses were made about the operation of these concepts and about methods for testing these concepts against the real world. When methods for observing and testing these concepts improved, the confidence level of some of the concepts increased, whereas confidence in other concepts decreased and these concepts died. Those concepts with higher confidence levels were developed into theories, such as the Copernican theory. As new ideas were introduced to existing theories, these theories were tested and found to accurately represent the real world. Kepler's work tested the theories of Copernicus with massive amounts of data—resulting in the development of principles with high confidence levels. These principles are Kepler's laws. The ideas of Newton regarding universal gravity provided a logical basis and rationale for Kepler's laws and further increased the levels of confidence in the laws.

6.2.3. Chemistry

The model for the evolution of the quantitative sciences, as presented in Chapter 1, does not treat the major science of chemistry. It was omitted from this model because there is not a single defining quantitative measure associated with chemistry. However, living systems cannot be treated without considering the evolution of chemistry. Therefore, a short treatment of the evolution of chemistry is in order.

As stated above, chemistry is the science that is concerned with the composition and structure of substances and the changes of composition and structure which they undergo. Chemistry is usually considered in two classes: analytical chemistry deals with the methods for separation of purer substances from mixtures, and of elements from compounds; synthetic chemistry treats methods by which complex bodies may be built up from simpler substances. In each case it is the changes of composition that concern the chemist. The theoretical study of chemistry is usually divided into three main branches: inorganic chemistry, organic chemistry, and physical chemistry.

The roots of chemistry, in its most rudimentary form, extend back to the dawn of humankind. Early humans had to distinguish between those substances which could be eaten, i.e., metabolized, and those which could not. They selected substances which were good for making tools based on characteristics such as ease of forming or hardness. Burning of materials to cause fire was a major concern of chemists for a long time and resulted in many chemical findings. Certainly primitive man was familiar with fire, as evidenced by ancient artifacts, and must have understood which substances would burn and the conditions under which these substances would burn. However, the artifacts from these early times of humankind are inadequate for identifying the processes and procedures used in the evolution of chemistry.

The science of chemistry had its origin in Egypt. Chemistry emerged from (1) the practical experience of workers in metals, glass, and pottery, and of use of dyeing and tanning materials, and (2) Greek and Eastern speculation on the nature of the material world. The school founded in Alexandria was the natural meeting place of the two streams—from their union came the alchemy of the Arab conquerors, the iatrochemistry of the medical chemist, and finally our modern science. In all the older theories of the origin of the universe, there was the idea that there was some primordial element or principle from which the visible universe was derived. Water, air, fire, and earth were each thought by various factions to be the primal element. That all four were primal elements, and that the varieties of matter were made from their intermixture, was the concept of Empedocles, who regarded each element as distinct and unchangeable. But the doctrine of the four elements that gave so powerful an impulse and direction to chemistry was that taught by Aristotle. The importance of his doctrine lay not so much in the nature he assigned to matter but in the broad principle that one kind of matter could be changed into another kind—in a word, that transmutation was possible. The idea was that underlying all tangible bodies was an indeterminate matter-stuff whose properties give matter its particular form.

Alchemy, the process of transmuting the common into the precious, and iatrochemistry, the science of the application of chemistry to medicine, were the early thrusts of chemistry. Alchemy and iatrochemistry were the mainstream of chemistry and were the basis for the birth of modern chemistry in the late 1600s. Much of early chemistry can probably be typified by many ideas with very low confidence levels. One of the main thrusts of early chemistry, alchemy, was an idea with such a low confidence level that it subsequently died.

Modern chemistry is said to have begun with Robert Boyle. He established by experiments the law known by his name—that the volume of a given mass of air varies inversely as the pressure upon the gas. Boyle also brought into question the long-held belief of the alchemists and the iatrochemists in *tria prima*—a belief that salt, sulfur, and mercury were the primary elements of chemistry. He also performed noteworthy experiments on the nature of fire. Boyle's idea was the first one associated with chemistry that proceeded to the law stage of science evolution.

Inquiry into the nature of fire occupied the attention of chemists for a considerable time and resulted in significant advances in chemistry. There were false starts, such as the idea and theory of phlogiston. However, the discoveries of oxygen and the law of the conservation of mass in chemical reactions resulted from these inquiries into the nature of fire. These discoveries were due mainly to the work of Lavoisier and his research on combustion that he started in 1772. The unique features of his work were the use of the balance and his way of interpreting quantitative results to show that no measurable matter disappears in any chemical change. The law of the conservation of mass enunciated by Lavoisier is a fundamental principle of quantitative chemistry.

A next major step in the evolution of chemistry was the atomic theory. This theory resulted, in the main, from the ideas and work of John Dalton, the results of which were published in 1808 in Dalton's *New System of Chemical Philosophy*. The atomic theory was further refined by A. Avogadro, who drew a distinction between the atom, the ultimate particle of an element that can take part in a chemical change, and the ultimate particle of an element of compound that can exist by itself; the latter he called a molecule. Avogadro's hypothesis became the very cornerstone of modern chemistry. Every advance after that time in one way or another provided fresh evidence of its truth, so much so that it became customary to refer to Avogadro's law. It is important to note that it took nearly a half century for Avogadro's ideas to secure general recognition among chemists. In essence, the ground work for quantitative relationships for chemical changes had been established.

The next step, which resulted in the establishment of a completely logical system of chemistry with consistent formulas and atomic weights, was taken by Cannizzaro in 1858. He insisted upon the necessity of accepting Avogadro's hypothesis as universally applicable. Cannizzaro was able to deduce correct values for atomic weights by showing that, once the molecular weight of a compound had been established by a determination of its vapor density, the value of the atomic weight of some particular element in the compound must be fixed within a certain limit.

Relationships between the values of atomic weights of elements were next to emerge. Correlation of the properties of elements with their atomic weights resulted in J.A.R. Newlands' law of octaves in 1866 and the periodic system of classification by D.I. Mendeleyev in 1869.

From these early beginnings, chemistry has evolved into three main branches: inorganic, organic, and physical chemistry. Inorganic chemistry is concerned with the study of the elements and compounds other than those of carbon—with a few minor exceptions—and was developed primarily through the investigations of minerals. Organic chemistry is concerned with the study of carbon compound substances of plant or animal origin. Although it was recognized that the two classes of substances were amenable to the same general chemical laws, it was supposed that a fundamental difference existed between the mineral substances and those which were the products of life processes. However, during the middle of the nineteenth century it was found that organic substances could be prepared from their elements or from inorganic compounds. It might seem strange that a single element (carbon) and its compounds should constitute the basis for one of the major branches of chemical science. However, when the relative volumes of the subject matter are considered the division appears reasonable because the known compounds of carbon far outnumber the noncarbonaceous compounds of all the other elements combined. Physical chemistry covers the ill-defined territory between physics and chemistry, as biochemistry does between chemistry and biology. A physical chemist is one who observes natural phenomena and records his observations, but maintains a judicious balance between the material changes themselves, the physical effects which accompany them, and the mathematical machinery by means of which the whole can be most compactly condensed.

Each of the three main branches of chemistry has reached the quantitative stage of evolution with formal laws and a high confidence level.

The evolution of chemistry further demonstrates the validity of the model for the evolution of the quantitative sciences. Chemistry started with a number of ideas with low confidence levels, proceeded with some ideas reaching the concept stage, then the theory stage, and finally validated sufficiently well against the real world to be called laws.

6.2.4. Other Factors

Two conditions are necessary for the evolution from fundamental principles to a quantitative science. The principles and the science must be applicable to problems of importance to humankind, and there must be a change in the way people think about the subject—from the old way to a new way based on the fundamental principals and evolving science. The fundamental principles for the nonliving systems quantitative sciences were all used to evolve applications which were important to humankind. It is anticipated that a quantitative living systems science will evolve in a similar way. A historical perspective of the evolution of the quantitative nonliving systems sciences from fundamental principles to acceptance and application of these principles provides a model for the further evolution of a living systems science. Thomas S. Kuhn (1970), a philosopher and science historian, provides such a historical model in his book *The Structure of Scientific Revolutions*. Kuhn's thesis is that at a particular point in time there are existing *ways of thinking*, i.e., paradigms, which must be overcome by a new concept before this new concept can be accepted. In Kuhn's view, there must be a scientific revolution before an old paradigm is overthrown by a new paradigm. He states that "probably the single most prevalent claim advanced by the proponents of a new paradigm is that they can solve the problems that have led the old one to a crisis. When it can legitimately be made, this claim is often the most effective one possible."

Kuhn continues:

> Claims of this sort are particularly likely to succeed if the new paradigm displays a quantitative precision strikingly better than its older competitor. The quantitative superiority of Kepler's Rudolphine table to all those computed from the Ptolemaic theory was a major factor in the conversion of astronomers to Copernicanism. Newton's success in predicting quantitative astronomical observations was probably the single most important reason for his theory's triumph over its more reasonable but uniformly qualitative competitors.

Kuhn's historical perspective of the evolution of science provides a good model for the acceptance and growth of a quantitative science. It can be used as a model for changing current paradigms of living systems behaviors to a new paradigm based on the quantification of these behaviors. The development of this new paradigm starts with this chapter which presents the characteristics of animals from the view that their behaviors can be quantified. The current paradigm is based, in the main, on qualitative descriptions of behavioral characteristics. The new paradigm shows how these qualitative descriptions can be converted to quantitative parameters using the fundamental principles developed in the preceding chapters.

6.3. Status of Living Systems Science Evolution

To proceed to the next stage in the evolution of a quantitative living systems science, it is necessary to determine where the science is in its evolution. The model described above was used to determine the status of the living systems science.

Following the evolution model, it is first necessary to give a very broad definition of the subject of the science. The definition of living systems that has gained acceptance in the systems sciences is *the science of all things living, ranging from the smallest living things (cells) to the largest living things (large groups of animals and humans)*. This definition, as used by systems scientists, is meant to be somewhat more inclusive than biology, which is defined as the science of life.

6.3.1. Ancient Beginnings

The next step in the model is to use available data to form some understanding of the beginning roots of the science. The artifacts of long ago that bear on humankind's understanding of living things are, at best, rare. There are few ideas that can be deduced from these artifacts. The confidence levels associated with ideas generated from these artifacts are extremely low. Another source of data is the primitive cultures that still exist, for example, the cultures of the aborigines of Australia. From these data, it can be deduced that primitive humans were interested in the characteristics and behaviors of plants and animals which were important to obtaining food. Otherwise, how would they have been able to gather proper types of plants for food and to successfully hunt animals? Also, it was important for humans to understand the characteristics and behaviors of predators of humankind in order to develop survival behaviors. It can be concluded that primitive man must have had a body of living system knowledge that was handed down from generation to generation.

The artifacts, pictures, and carvings from ancient civilizations give credence that the ancients had an empirical understanding of living systems. An Assyrian bas-relief shows that hand-pollination of the date palm was practiced at least nine centuries before Christ, and there are records that this palm was cultivated as long ago as 6000 B.C. The ancients must have known that there were two kinds of date palm trees and that both were necessary for the production of fruit, though nothing was known of the mechanism of breeding at that time. Sculptures and drawings also show that ancient Egyptians and Assyrians reared horses and cattle, indicating they knew something about selection. For example, an engraving on bone in an excavation in Mesopotamia, dating from about 2800 B.C., seems to be a pedigree chart of horses of several different types. The Chinese were cultivating rice as early as 5,000 years ago, and barley grains have been found in the tombs of mummies who lived about 5000 B.C. In ancient Babylon, clay models of various parts of the human body and ancient Babylonian writing show that some progress had been made in medicine. The practice of embalming the dead in ancient Egypt reveals that they had a knowledge of human anatomy. From all these facts, it is clear that the existence of a considerable body of living systems knowledge was possessed by the ancient peoples.

The term *living systems* is relatively new, having been coined in this century (Miller, 1978). The word *biology*, introduced by Treviranus (1776–1837), has been used extensively to cover all those studies dealing with the structure, nature, and behavior of living beings. The evolution of living systems science is found in the history of biology.

6.3.2. Greek and Roman Biology

Biology, like other sciences, can be traced to the Greeks. The earliest zoological writings of importance were produced by the Greeks. These early writers were philosophers who developed the deductive method of reasoning. These Greek writers refer to "the ancients," which seems to indicate that they knew of the body of philosophical and biological knowledge described above. Some of the more important men of the early Greek period and their ideas follow.

Thales (640–546 B.C.) believed that all life originated in the ocean. This idea is not dead even today. Anaximander (611–547 B.C.) believed in the spontaneous generation of life and in the idea that animals developed in the sea, becoming transformed into terrestrial animals. Alemaeon of Croton (c. 500 B.C.) described investigations of the structure and habits of animals. He described the optic nerves and tubes that lead from the ear to the nose, and he began the study of the development of the embryo. Empedocles (495–435 B.C.) also believed in the idea of spontaneous generation and held the belief that parts of organisms were kept separated by hate and were later brought together by love. But his more influential idea was that the blood is the seat of the innate heat. This led to the consideration of the heart as the center of the vascular system and chief organ of the *pneuma* which was distributed by the blood vessels.

Hippocrates (460–370 B.C.), sometimes called the father of medicine, made important contributions through his treatises *On the Sacred Disease* and *On the Nature of Man*. These treatises provide evidence of early attempts to make a classification of animals. Many of these treatises deal with anatomy and physiology. *On the Nature of Man* contains the doctrine of the four humors—blood, phlegm, black bile, and yellow bile—that make up the living body. This doctrine persisted until modern times, and traces of it remain in recent language and thought. The early Greek period came to a close with *Democritus* (460–357 B.C.). His ideas contained a materialistic view of the universe, including the idea that the soul perished with the body. He believed that the brain is an organ of thought and that types of animals might be distinguished by the quality of their blood.

The evolution of biology, as given above, is at best fragmentary and inadequate to permit adequate reconstruction. It was not until the time of Aristotle

(384–322 B.C.) that a complete biological work emerged and survived. Among Aristotle's surviving works, four stand out. Their subjects are (1) the psyche, that is, the soul or living principle; (2) histories about animals; (3) the generation of animals; and (4) the parts of animals. Aristotle used the term *psyche* for the principle that differentiates living from nonliving things, and he concluded that there were different kinds or orders of psyche or soul. These he considered to be the vegetative soul, the animal soul, and the rational soul. His were the first scientific treatises that had an influence upon modern biological thought—so good were his observations and deductions that they were considered authoritative for 20 centuries. Aristotle did not believe in special creation, but he did believe in an "internal perfecting tendency," or *intelligent design*. He was mainly interested in animals and was familiar with the facts of comparative anatomy, physiology, and embryology.

Aristotle's pupil Theophrastus (370–287 B.C.) was the first to study plants scientifically and is often considered to be the founder of the science of botany. Theophrastus understood the value of developmental study. He said that "a plant has the power of germination in all its parts for it has life in them all, wherefore we should regard them, not for what they are, but for what they are becoming." His works are valuable as perhaps the most complete biological treatises that have come down to us from antiquity.

Pliny the Elder (A.D. 23–79) compiled 37 volumes of natural history which remained a source of information regarding natural history for 15 centuries. These books document the folklore and superstitions of the time and, from our point of view, were ideas with very low confidence levels.

Dioscorides (A.D. ?–40) was a writer whose work had most influence on the course of botany, and in particular on botanical terminology. His work on plants was useful in medicine and is essentially a drug collector's manual. His work was pragmatic and almost devoid of general ideas.

The only other important person in the biological thought of antiquity is Galen (A.D. 130–200). He was a physician, the last of the great biologists of ancient times. His scientific works are among the most influential of all time. His books on anatomy were the standard for use in medical schools for about 15 centuries. His main idea was a physiological scheme which involved three kinds of pneuma or spirit in addition to air—*natural, vital,* and *animal spirits.*

After Galen and with the fall of the Roman Empire (A.D. 400), ideas stopped in biology. This period is usually designated as the Dark Ages and lasted for about a thousand years. The Church controlled all learning and all biological questions were referred to the ancient authorities. From the perspective of our model, there was no further evolution of living systems science during the Dark Ages.

6.3.3. Medieval Biology

From about A.D. 1350 through the sixteenth century, or the Renaissance peri-od, there was a gradual return to personal observation. The confidence level of biological ideas could now be tested against the "real world" through the process of observation and experiment. One of the most important men of the Renais-sance was Frances Bacon (1561–1626). He was perhaps responsible for promot-ing the process of observation and experiment.

6.3.4. Post-Sixteenth Century Biology

The return to first-hand observation, especially through dissection of the human body, led to comparative anatomy. The invention of the microscope led to microscopic anatomy and to an understanding of the structures of plants and ani-mals. Development of the microscope and understanding of anatomy led to many ideas about how living systems developed. Hieronymus Fabrizio (1537–1619) observed the development of the chick and began the establishment of the science of embryology. Casper Frederich Wolff (1733–1794) was the first to compare plant and animal development. His major contribution was to introduce the theo-ry of epigenesis—development involving gradual diversification and differentia-tion of an initially undifferentiated entity, such as a spore—as a replacement of the preformation theory.

First-hand observation and experimentation had a major impact on the gen-eration of ideas and the conformation, or refutation, of these ideas. The founda-tion for modern physiology was laid by the discovery, through observation and experimentation, of the circulation of the blood. William Harvey, in 1628, and Johannes Muller some 200 years later made significant discoveries regarding the circulation of blood. Observations made possible by the microscope resulted in the emergence of microbiology as a field of biology. There were many early suc-cesses in microbiology. Louis Pasteur (1822–1896) proved that microorganisms cause fermentation and Robert Koch (1843–1906) discovered the microorganism that caused cholera.

A major evolution in biology is the theory of evolution itself. This theory, like so many other scientific theories, had its genesis in the work of the ancient Greeks. This work included the idea that species change or evolve. During medieval times, the majority of biologists believed in special creation and the immutability of species. Charles Darwin's grandfather, Erasmus Darwin (1731–1802) expressed his views on evolution in his book *Zoonomia*. One of his beliefs was the inheritance of acquired characteristics. The most famous advocate of the theory that acquired characters are inherited was Jean Baptiste Lamarck (1744–1829). His theory of organic evolution was the most complete up to that time. Most authorities agree that the publication in 1859 of *The Origin of Species by Means of Natural Selection* by Charles Darwin (1809–1882) had a major impact upon the thinking of that time. Darwin's ideas of natural selection and organic evolution were the outgrowth of the ideas of his grandfather and other men. Charles Darwin collected the data that provided a high degree of confidence in these ideas and elevated them to a theory.

The evolution of cell theory was made possible by greatly improved microscopes and delicate techniques and processes. Cell theory was formulated in the publications of Matthias Jocob Schleiden (1804–1881) and Theodor Schwann (1810–1882). A basic principle of biology is attributed to Max Schultze (1825–1874). Who established through his research that protoplasm is the fundamental substance of both plants and animals.

The development of biology after the fifteenth century was both rapid and diverse. A number of fields of biology evolved, such as anatomy and embryology. One field was related to the relationships of plants and animals and of the species within the plant and the animal kingdoms. This field is systematic biology. Carolus Linnaeus (1707–1778) was a leading scientist in this field. His idea was to describe and classify all the existing species of plants and animals using a binomial system of nomenclature. This system of nomenclature gained wide acceptance and is in use today.

6.3.5. Modern Biology

By the start of the twentieth century, a number of ideas had evolved to the confidence levels associated with fundamental principles, doctrine, and laws. Brief descriptions of these ideas follow:

1. Living matter consists of protoplasm. This property of matter separates living from nonliving systems.
2. The smallest amount and organization of living matter that can exist and exhibit the properties of life is the cell. The basic unit of organization of living things is the cell, and all organisms are composed of cells.
3. The processes of nutrition and of respiration are fundamentally the same for all living things.
4. All living things are the product of living things. This doctrine is essentially equivalent to the physics doctrines of the conservation of energy and mass.
5. The fundamental mode of reproduction of animals and of plants is essentially identical.
6. Classification systems have been developed for the identification of living systems and of their similarities and differences.
7. Mendel's ideas on the most basic patterns of inheritance in sexually reproducing organisms has become Mendel's laws.

Both living and nonliving systems sciences developed rapidly in the twentieth century. More powerful microscopes, including the electron microscope, provided a means for observing living systems phenomena of smaller and smaller amounts of living matter. In addition, more refined and powerful experimental processes and procedures evolved and led to many important ideas and discoveries.

One of the greatest triumphs of twentieth-century biology has been the elucidation of the mechanics underlying the basic Mendelian patterns of inheritance. Although Mendel devised a powerful explanation for the mechanisms underlying the patterns of inheritance, this explanation was a concept with rather low confidence levels. Geneticists, biochemists, biophysicists, and molecular biologists, with their vastly improved instruments, tools, processes, and procedures converted Mendel's concept into biological dogma. The central dogma of molecular biology is one of the most important concepts to have emerged to explain how genes make polypeptide chains. The central dogma is simply the idea that DNA codes for the production of RNA, RNA codes for the production of protein, and protein does not code for the production of protein, RNA, or DNA.

The idea of genetic information is central to the dogma of molecular biology. This idea is (1) there is a flow of information among the various classes of macromolecules (giant molecules) in living systems, (2) the source of the information is DNA, (3) within the structure of DNA molecules lies the necessary information to dictate the structures of the many different proteins in an organism, and (4) transmitting the information in DNA to proteins is the task of various types of RNAs. Francis Crick contributed two key ideas to the development of the central dogma and to genetic information. The first solved a difficult problem of explaining the relationship between a specific nucleotide (the basic chemical unit in a nucleic acid) sequence in DNA and a specific amino acid sequence in protein. There is no chemical affinity of nucleotides for amino acids. Secondly, Crick proposed a clever idea when he hypothesized that there is an adapter molecule that carries a specific amino acid at one end and that recognizes some sequence of nucleotides with its other end. Without being aware of Crick's "adapter hypothesis," other molecular biologists found and characterized the adapter molecules. These are small RNAs called transfer RNAs, or tRNAs. They recognize the genetic message and simultaneously carry specific amino acids, thus translating the language of DNA into the language of proteins.

Another problem existed: How did the genetic information get from the nucleus to the cytoplasm? The large bulk of DNA of a eukaryotic cell is confined to the nucleus, but protein synthesis is carried out in the cytoplasm. Crick, together with the South African geneticist Sydney Brenner and the French molecular biologist Francois Jacob, tackled this problem with another key idea, the "messenger hypothesis." According to the messenger hypothesis, a specific type of RNA molecule forms as a complementary copy of one strand of the gene. It contains the information from that gene, so there should be as many different messengers as there are different genes. This messenger RNA, or mRNA, then travels from the nucleus to the cytoplasm. There it serves as a template on which the tRNA "adapters" line up to bring amino acids in the proper order into a growing polypeptide chain in the process called translation.

In summary, a given gene is transcribed to produce a messenger RNA complementary to one of the DNA strands, and transfer RNA molecules translate the sequence of bases to the mRNA into the appropriate sequence of amino acids. This process, in essence, describes the generation of genetic information, the transfer of this information, and the use of this information to cause a precise behavior of living matter.

The concept of chemical information matured in the twentieth century from its early beginnings in the mid-1800s. This concept is treated in Chapter 4.

The idea of neural information was the first of the three ideas of biological information to emerge because it is the most readily observable. The idea of neural information resulted from the work of Luigi Galvani which is described briefly in Chapter 4. Galvani began his investigations on the action of electricity upon muscles of frogs in about 1771 and enunciated his theory of animal electricity in 1791. Galvani's observation that the suspension of frogs on an iron railing by copper hooks caused twitching in the muscle of their legs. This led to the invention of his metallic arc. He constructed his arc of two different metals. When he placed one metal in contact with the frog's nerve and the other one with a muscle, he observed the contraction of the muscle. In Galvani's view, the motions of the muscle in his experiments were the result of the union, by means of the metallic arc, of its exterior or negative electrical charge with positive electricity which proceeded along the nerve from its inner substance. By the end of the nineteenth century, much had been learned about the neural system, its physiology, and its functions. In the present century, the concept of neural information and its association with observable behavior is well established.

By the second half of this century, the concept of information had become well-established in living systems science. James G. Miller, in his book *Living Systems*, differentiates living from nonliving systems, in part, by their ability to process information. Living systems are processors of information whereas nonliving systems are not.

It is probably fair to say that the concept of biological information is one of the most fundamental and important to evolve during this century. Information is a fundamental characteristic and a determinant of living systems and their behaviors.

6.4. The Next Phase

The models for the evolution of science and for the quantification of science provide guidance for the next phase in the evolution of living systems science. The nonliving systems sciences evolved from descriptive sciences to quantitative sciences with the quantification of the fundamental parameters of the science. Further evolution of the nonliving sciences was achieved by the application of the quantitative concepts of the sciences.

Living systems science is in the descriptive and qualitative stage of evolution. The next phase of evolution to a quantitative living systems science is to test and validate the quantitative living systems science principles listed at the beginning of this chapter. The remaining chapters of this volume test the principles of quantitative living systems science.

6.5. Summary

The sciences have evolved over the course of humankind's existence in a somewhat consistent pattern of stages. A particular science may be in one stage of evolution while another science may be in another stage. This pattern for the evolution of science has been developed in this chapter based on the evolution of the physical sciences because the latter have evolved further than other sciences. The science evolution pattern is based on the emergence of ideas and the development of ideas to the extent that they have been verified as invariant and hence are laws of science. This pattern is shown to be applicable to the evolution of living systems science and therefore can be used as a model for the further evolution of quantitative living systems science.

The model is used to describe the stage of living systems science evolution. The major ideas, concepts, and laws of living systems science are described so they can be used as a base for the further evolution of the science. The next stage of the evolution of living systems science is identified.

CHAPTER 7

Autonomous Animal Behavior

7.1. Introduction

The evolution of science model described in Chapter 6 provides guidance for testing, verifying, and applying the fundamental principles of quantitative living systems science. We need to (1) identify the criteria for selection of the systems to be tested, (2) synthesize system concepts, (3) identify the hypotheses and theories to be verified, and (4) identify and select applicable data that has been verified to a high level of confidence.

There are criteria for selecting living systems to be used for testing and verifying the fundamental principles of a quantitative living systems science: (1) the characteristics of the system must be readily observable, (2) the system's characteristics must be stable, (3) a large number of these systems must exist, and (4) applicable data on the system's characteristics must be readily available.

Based on these considerations and criteria, animals were selected as the living systems to investigate. Animals are important to humankind. The behaviors of animals are relatively easy to observe due to their rapid motion. There is a wealth of data regarding animals, and many ideas and concepts concerning animals have been validated. Further, the focus is on mechanical forces because these forces are used by animals to change shape and to move—which are the bases for just about all readily observable animal behavior. We focus on neural information because it causes these mechanical forces and mechanical work.

The autonomous behaviors of animals are those most constant over time and the most consistent of all behaviors. These are the behaviors of breathing, blood circulation, and the like which are caused by information generated inside the body of an animal. Therefore, as indicated by the model, these behaviors provide a good place to start in testing the concept of behavioral information and behavior. In addition, there is now a robust literature on autonomous behaviors and the causes of these behaviors. This robust literature include materials readily available to most people—such as encyclopedias, biology textbooks and general books on the subject.

The concepts developed in preceding chapters, along with the extant data on autonomous behaviors, were used to synthesize concepts of behavior and information as they relate to the autonomous behaviors of animals. These concepts were then tested against validated data in order to establish the level of confidence of the concepts. The concepts were developed and based on readily available data, and test data were also selected based on its availability. Generally accepted and readily available data were selected in order to minimize the need for a large number of references.

After validating the general concepts for the autonomous behaviors, these concepts were further defined and detailed.

7.2. Concepts

The fundamental concept of quantitative living systems science comprises the following principles: (1) behavioral information is a fundamental determinant of behavior and it can be quantified, (2) behavior can be quantified in terms of the energy in the behavior, (3) living systems have an innate capacity to direct energy based on their structure and organization, and (4) there is a fundamental relationship among behavior, the capacity to direct energy, available energy, and information. Further, the concept is that these four principles form the basis for quantitative living systems science.

This fundamental concept must be further developed and specified in order to construct a concept of the autonomous behaviors of animals that can be tested using existing data. The concept development started with the obvious autonomous behavior of our heartbeat, breathing, and digestion. For example, breathing behavior can be readily observed by each of us, and it has been recognized for a long time as a behavior vital to life itself. The earliest writings of the ancients describe the "breath of life," which is an indication that they observed this behavior and its necessity for life. The autonomous blood circulation behavior was discovered much later by William Harvey (1628). The mechanisms for breathing and blood circulation were not well understood until the discovery of the workings of contractile tissue. It is now understood that these autonomous behaviors are associated with metabolism, that the fundamental mechanism of metabolism is the same for all living systems, and that fundamental metabolism is at the cell level. It is an established fact that living cells must have a continuous supply of oxygen and glucose for them to sustain life. It is also an established fact that animals do not store oxygen and therefore must continuously obtain oxygen from their environment. From these facts, we can construct a conceptual model of the mechanism by which the autonomous behaviors arise.

These fundamental facts provide a basis for a concept of an autonomous behavior. First, cells and organisms with very little organization can obtain oxygen and nutrients directly from the environment. However, as soon as cells organize into structures where individual cells cannot be supplied oxygen and nutrients directly from their environment, there must be a structure which will allow those cells not in direct contact with the environment to obtain oxygen and nutri-ents. Simple organisms such as the sponge fulfill the vital function of supplying oxygen and nutrients to their cells by moving liquids and particles over these cells with tiny hair-like appendages of some cells. These hair-like appendages are called cilia. These special cells with cilia have a capacity to contract, thereby pro-viding motion to the cilia which either move the animal or move liquids within the animal. Large animals, such as humans, have well-developed lungs, hearts, circulation systems, and digestive systems that continuously provide nutrients and oxygen to each individual cell in our bodies. Without this continuous supply of nutrients and oxygen, the cells die, which, in turn, results in the death of the organism composed of these cells.

7.2.1. Heart and Circulation

Our heart and circulation system continuously provide each of our cells with fluids which contain nutrients and oxygen. The heart pumps oxygen and nutrient-rich blood to all our cells through the circulation system, where the oxygen and nutrients are transferred to the cells and the byproducts of cell metabolism are transferred to the blood. The oxygen and nutrient-depleted blood are then returned to the heart. Oxygen-depleted blood is then pumped to the lungs where the waste gases are expelled and oxygen is replenished in the blood. From our point of view, the heart has a most important characteristic—it continues to beat even when it is removed from the body. This characteristic is based on the heart having an information generator of its own—the pacemaker. This information generator can operate on its own without information from other sources within the body and it causes the heart to beat at a regular rate.

Based on the pacemaker characteristics of the heart and the heart's other characteristics, a behavioral concept that is an elaboration of the concepts of behavior given in Chapter 5 was synthesized. As described in Chapter 5 behavior can be quantified in terms of the energy in the behavior, the capacity of a living system to direct energy, available energy, and behavioral information. However, we did not address the question of where the behavioral information came from and how it originated or was generated. In the human heart, the behavioral infor-mation of the pacemaker originates in the heart itself and is generated by a known chemo-electric process which results in an action potential (behavioral informa-tion quantum). Some heart muscle fibers are specialized for the pacemaking func-tion; they initiate the rhythmic contraction of the heart. This pacemaking function is due to a unique class of potassium ion channels found in cardiac muscle fibers. These channels tend to remain somewhat open following an action potential, but they gradually close. As they close, the cell becomes less negative and eventually reaches threshold, thereby initiating the next action potential. Figure 7.1 depicts an idealized concept of the relationship for this basic heart behavior.

Each time a new quantum of information is generated and transmitted to the heart muscle fibers, they contract based on their capacity to direct energy—using some of the available energy—and cause the heart muscle's contraction behavior. Each of the fundamental parameters (i.e., capacity to direct energy, available energy, behavioral information, and behavior) depicted in Figure 7.1 can be measured and quantified. This conceptual model of the heart represents the elemental parameters of the heart for one information generator, namely, the pacemaker. When there are other information generators that can modify the heart rate by influencing the pacemaker information generator, the model becomes more complex. However, for the purposes of establishing a fundamental concept of quantitative autonomous behaviors, these other factors do not need to be considered at this time.

7.2.2. Breathing

Breathing, as mentioned earlier, is one of the most obvious of our autonomous behaviors. In addition, our breathing must be ongoing in order for a continuous supply of oxygen to be available for each cell in our body, thereby keeping us alive. Breathing is one attribute of the gas exchange system that allows us to breath in air, extract oxygen from this air, transport oxygen to our cells, transfer oxygen to cells, transfer carbon dioxide from cells, transport carbon dioxide to the lungs, and expel carbon dioxide into the environment. The part of this gas exchange system treated below is the most observable of the breathing behavior—the mechanics of ventilation.

Human breathing behavior involves the lungs, the thoracic cavity, the diaphragm, the pleural membranes, and the muscles in the chest, as shown in Figure 7.2. The lungs are suspended in the thoracic cavity, which is bounded on the top by the shoulder girdle, on the sides by the rib cage, and on the bottom by a domed sheet of muscle (the diaphragm). The thoracic cavity is lined on the inside

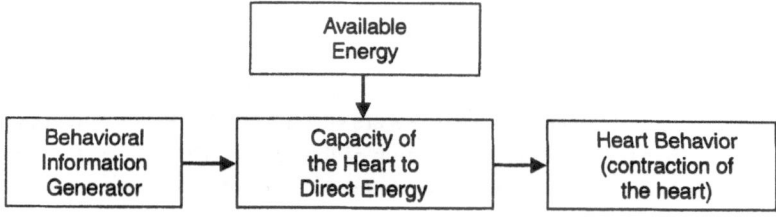

Figure 7.1. Idealized concept of heart behavior.

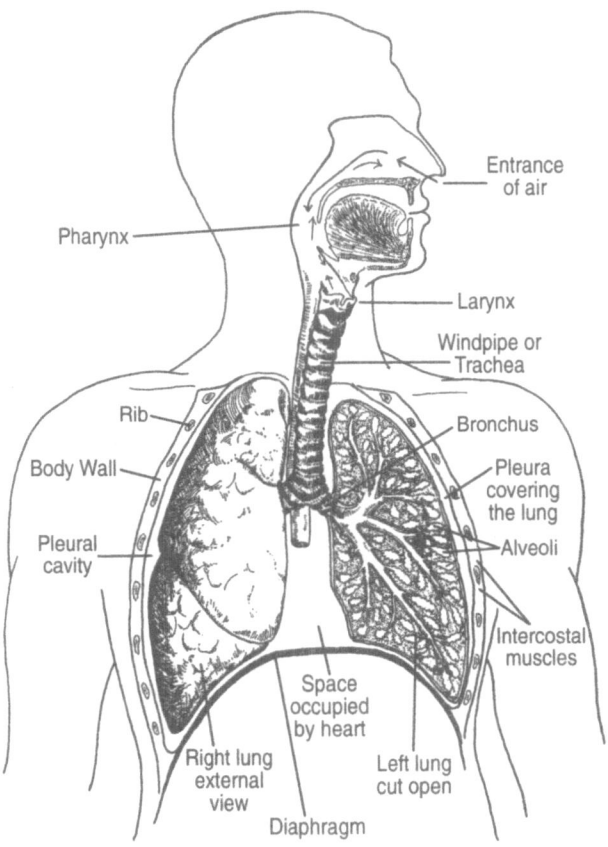

Figure 7.2. The human respiratory system.

by pleural membranes, which divide it into right and left pleural cavities. The pleural cavities are closed spaces, so any effort to increase their volume will create negative pressure, that is, a suction, within them. Negative pressure within the pleural cavities causes the lungs to expand as air flows into them from the outside. This is the mechanism of inhalation. The diaphragm contracts to begin an inhalation. This contraction pulls the diaphragm down, increasing the volume of the thoracic and pleural cavities. As pressure in the pleural cavities becomes more negative, air enters the lungs. Exhaling begins when the contraction of the diaphragm ceases, the diaphragm relaxes, and the elastic recoil of the lungs pushes air out through the airways. The volume of the thoracic cavity can also be increased by the intercostal muscles between the ribs expanding the thoracic cavity by lifting the ribs up and outward.

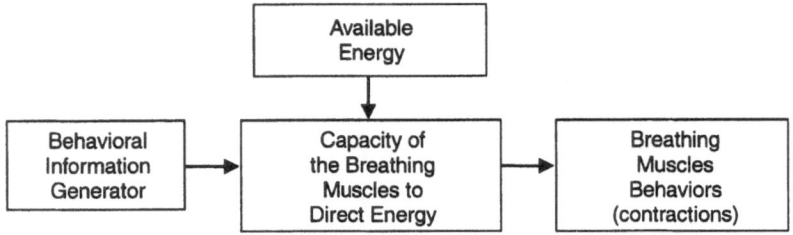

Figure 7.3. Idealized concept of breathing behavior.

The muscles associated with breathing, just as other muscles described previously, require information for their contractile behaviors. The information that
causes contraction of the breathing muscles is generated in our brain stem just
above the spinal cord by the medulla. Figure 7.3 depicts an idealized concept of
the relationships for the breathing behavior.

The actual breathing system is more complex than the simple concept
depicted above. For example, chemosensors (information generators) on the heart
blood vessels and on the surface of the medulla generate information which is
sent to the information generator in the medulla, where it is combined to change
the information generated in the medulla. The chemosensors on the large blood
vessels leaving the heart are sensitive to the partial pressure of oxygen in the
blood. The chemosensors on the surface of the medulla are sensitive to the partial pressure of carbon dioxide in the blood. These chemosensors cause changes
in the information generated in the medulla in order to match the breathing rate
with changing metabolic needs of the cells.

The idealized breathing concept depicted in Figure 7.3 is adequate to show
how this behavior relates to the general concept of the relationships among the
parameters of behavior given in Chapter 5.

7.2.3. Digestion

Another of our readily observable behaviors is the intake of food and the
elimination of the waste products of food. An important autonomous behavior
related to food processing and digestion is the movement of food and the byproducts of food digestion through our digestive tract.

Our digestive tract is a continuous tube that runs from mouth to anus and has
separate compartments that have specific functions. The cellular architecture of
the tube follows a common plan throughout. Four major layers of different cell
types form the wall of the tube, as shown in Figure 7.4. These layers differ from
compartment to compartment, but they are always present. Starting in the cavity,
or lumen, of the gut, the first tissue layer is the mucosa. Nutrients are absorbed

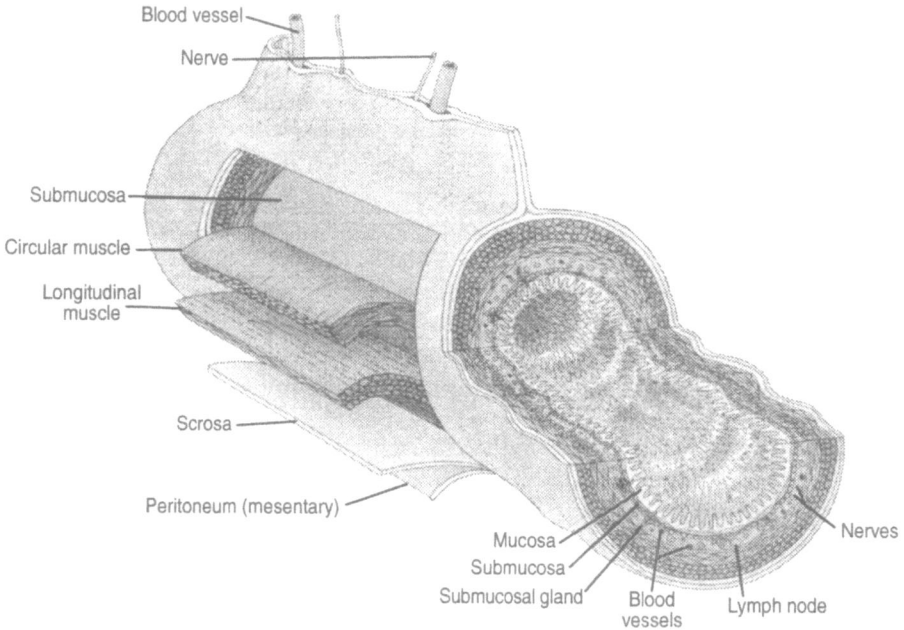

Figure 7.4. Tissue layers of the human gut.

across the membranes of the mucosal cells; in some regions of the gut, those membranes have many folds that increase their surface area. Mucosal cells also have secretory functions. Some secrete mucus that lubricates the food and protects the walls of the gut. Others secrete digestive enzymes, and still others in the stomach secrete hydrochloric acid. At the base of the mucosa are some muscle cells. Just outside the mucosa is a layer called the submucosa where the blood and lymph vessels are located that carry absorbed nutrients to the rest of the body. The submucosa also contains a network of sensory and regulatory neurons that control the various secretory functions of the gut.

External to the submucosa are two layers of smooth muscle cells responsible for the movements of the gut. The innermost layer has its cells oriented around the gut and is called the circular layer. It constricts the lumen. The outermost layer of muscle, the longitudinal layer, has its cells arranged along the length of the gut. When it contracts, the gut shortens. Between these two layers of muscle is a network of nerves that control the movements of the gut and coordinate the different regions with one another. Surrounding the gut is a fibrous coat called the serosa, which in turn is covered and supported by a tissue called the peritoneum.

Food entering the mouth is chewed and mixed with the secretions of salivary glands. The muscular tongue then pushes the mouthful, or bolus, of food toward the back of the mouth cavity. By making contact with the soft tissue at the back of the mouth, the food initiates a complex series of neural reflex actions commonly known as swallowing. The act of swallowing involves many muscles doing a variety of jobs that propel the food through the pharynx and into the esophagus without allowing any of it to enter the windpipe or nasal passages.

Once the food has reached the esophagus, peristalsis takes over and pushes the food toward the stomach. Peristalsis is a wave of smooth muscle contractions that move progressively down the gut from pharynx to anus. The smooth muscle of the gut contracts in response to being stretched. The act of swallowing a bolus of food stretches the upper end of the esophagus, and this initiates a wave of contraction that slowly progresses down the entire gut, pushing the contents of the gut toward the anus.

The movement of food from the stomach into the esophagus is normally prevented by a thick ring of circular smooth muscle at the esophageal–stomach junction. This ring of muscle, called a sphincter, is normally constricted. Waves of peristalsis cause it to relax enough to let food pass through. The pyloric sphincter is located between the stomach and the intestine and controls the flow from the stomach to the intestine. Passage of digestive tract byproducts from the lower end of the gut is controlled by the sphincter surrounding the anus.

The peristalsis behavior is controlled by neural information generation and distribution which is intrinsic to the digestive tract. This behavior is autonomous as it is not controlled by the central nervous system. A simple idealized concept of peristalsis behavior is depicted in Figure 7.5.

The simple model shown in this figure is for one elemental section of the gut and the start of another. An elemental section of gut starts with an information generator. Stretching of the gut causes the generation of information which is transmitted by neurons to the smooth muscle associated with these neurons. Upon receiving this information, the muscle contracts in such a manner to move the bolus of food down the gut. When the bolus of food enters the next elemental section of gut, it stretches the gut and the generator associated with this next ele-

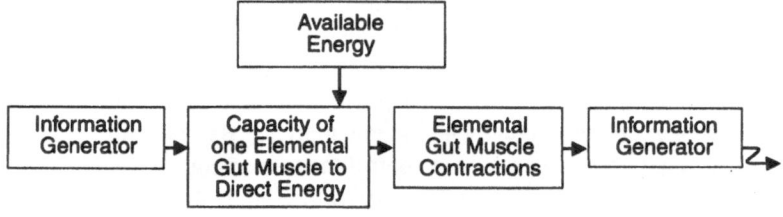

Figure 7.5. Idealized concept of digestive behavior.

mental section produces information that activates this element. This progressive contraction behavior of the gut passes food and the byproducts of digestion through the gut. This process is based on the generation of information, the capacity of the muscle in the gut to contract in its own special way based on the characteristics of the muscle, the availability of energy for use by the muscle, and the contraction behavior of the muscle.

7.2.4. Basic Concept

A basic concept emerges from the above considerations of our most obvious autonomous behaviors. First, these behaviors are associated with metabolism—one of the two key processes characteristic of all living systems. Second, the fundamental parameters of these behaviors are the local generation of information, the capacity to direct energy of the muscles associated with these behaviors, available energy, and the energies in the contraction of the muscles. Third, the information associated with these vital behaviors is generated internal to our bodies and is, for all practical purposes, independent of our central nervous system.

This basic concept for the autonomous behaviors was synthesized from the fundamental principles presented in earlier chapters and from existing descriptive characteristics of autonomous behaviors. Therefore, there is a high probability that the descriptive basic concept is correct. However, quantification of this basic concept was much less certain. The task was to increase the confidence level in a quantitative concept for the obvious autonomous behaviors.

It was hypothesized that existing descriptive and quantitative data on autonomous behaviors could be used to verify the quantitative autonomous behavior concept presented above. The autonomous behaviors of the cardiovascular system are treated first, followed by the respiration system, and then the digestive system.

7.3. The Cardiovascular System

Many invertebrates and all vertebrates have cardiovascular systems which consist of a muscular pump, or heart, that moves blood through a closed network of vessels. This system transports nutrients, respiratory gases, hormones, metabolic products and wastes, and elements of the immune system to the animal's cells.

It was hypothesized that the basic concepts of the heart and circulation system behaviors described above would allow these behaviors to be quantified and that there is sufficient data to test and verify the concepts. To test these hypotheses, it was necessary to (1) identify and specify the structure of the cardiovascular system of specific animals—it is this structure that determines a cardiovascular system's capacity to direct energy, (2) identify the information generation and distribution system which causes the cardiovascular behavior, (3) identify and specify the energy available for the behaviors, and (4) identify and specify the energy in the behaviors.

7.3.1. Cardiovascular Structure

There are seven major types of cardiovascular structures, ranging from the very simple to the extremely complex. They are, in increasing degrees of complexity: tubular, arthropod, mollusk, fish, amphibian, reptiles, birds, and mammals. A short review of each structure is given below.

7.3.1.1. Tubular. The common earthworm provides an excellent example of a primitive type of circulation. Down the back runs a tube filled with blood, and down this tube from the tail to the head pass regular waves of contraction moving at a rate of about half an inch per second. The waves of contraction squeeze the blood forward just as a thin rubber tube can be emptied by drawing a finger and thumb down it. Waves of contraction of this nature which pass along a structure are peristaltic waves. The blood is squeezed from the dorsal artery into arterioles, which connect with small veins; the veins empty into a ventral vein which returns the blood to the dorsal artery. Tubular hearts are the primitive type from which the more complex hearts evolved. Hearts of this type are found in worms and also in the ascidia. These are a primitive group which form a link between invertebrates and vertebrates.

7.3.1.2. Arthropod. The vast majority of living animals belong to the arthropod phylum which contains such classes as Crustacea, Arachnida, and Insecta. This phylum may be divided into two great groups: (1) the Crustacea, which breathe by gill, and (2) the spiders, scorpions, and insects, which are mostly land animals and which breathe partly or entirely by tracheae.

In the whole class of arthropods, the circulatory structure has certain features in common. The heart is surrounded by a venous sinus and the blood passes into the heart by means of holes, or ostia, provided with valves. When the heart contracts, the contraction closes the ostia and drives the blood out through the arteries. In most forms there is no system of capillaries and veins; instead, the organs are surrounded by venous sinuses.

The gill-breathing forms have fairly powerful hearts. The king crab *Limudus* is noteworthy because it has a better developed circulatory system than any other arthropod. This form is a primitive type of arthropod and is intermediate between Crustacea and spiders. It possesses a closed circulatory system of arteries, capillaries, and veins, and a very powerful heart. On the other hand, some of the smaller Crustacea have only very rudimentary circulatory mechanisms, without any true heart. The hearts of Crustacea are superior to those of other arthropods but are distinctly inferior to those of the mollusca; for example, a crab and an octopus of similar size are both cold-blooded, sea-living, gill-breathing animals of about equal activity, but the heart of the crab is a far simpler and weaker form of pump than is the heart of the octopus.

The tracheated arthropods include the spiders, scorpions, and insects, which have very simple and inefficient hearts. The typical heart is a tubular structure running along the dorsum and is divided into segments by valves. Waves of contraction pass forward, driving the blood either forward or out through lateral arteries. The reason for the imperfect development of the heart in these forms is that they are not dependent on the circulation for their supply of oxygen, but breathe through minute branching tubes or tracheae which carry air from the surface of the body directly to the cells. This is a very efficient and economical respiratory mechanism for animals up to a certain size, as is indicated by the success of the class Insecta which contains more than half of the known species of animals but in which the vast majority of species are less than one gram in weight.

7.3.1.3. Mollusk. In the molluscan circulatory system, the blood returning from the body is collected in auricles that contract and squeeze the blood into the ventricle which expels it into arteries. This organization contrasts with the arthropod heart in which the ventricle is suspended in a venous sinus. The contractile auricle ensures the proper filling of the ventricle and represents an important advance in the efficiency of the heart as a pump. The vertebrate heart is constructed on the same principle. In the highest mollusk, the cephalopods (e.g., octopus), the circulatory mechanism is complex for, in addition to the main heart, accessory hearts are provided to pump the blood through the gills. The difference between the efficiency of the arthropod and molluscan hearts is illustrated by the fact that in the lobster, which is a particularly active crustacean, the arterial pressure is only 12 cm of water, whereas in the octopus the arterial pressure may rise as high as 115 cm of water.

7.3.1.4. Fish. Fish descended from other sea-living forms. Their cardiovascular systems are more complex than earlier sea forms such as the mollusk. The cardiovascular system of an elasmobranch, such as the dogfish, is a good example of a primitive vertebrate cardiovascular system. The heart consists of four divisions: the sinus venosus, the auricle, the ventricle, and the bulbus arteriosus. Each of these chambers is divided by valves from its neighbors, and the heart functions by a wave of contraction which passes down from the sinus to the bulbus arteriosus and drives the blood before it from one chamber to another. This wave of contraction passes at a rate of about 4 inches per second. When one chamber contracts, a certain interval is necessary to allow the blood to pass into and distend the next chamber; because the wave of contraction is delayed for a fraction of a second between each of the chambers this time interval is ensured. The muscles forming all the chambers have the power to contract rhythmically without any stimulus being applied, but the natural rhythm of the sinus muscle is the highest and hence the sinus acts as a pacemaker to the rest of the heart.

On the whole, the hearts of the different classes of fish do not differ in any important respect. The lampreys are the most primitive group of vertebrates. Their hearts are slightly more primitive than the dogfish heart, but the general arrangement and properties are the same. In the leleosts there is no true bulbus arteriosus, but otherwise there is no important difference from the elasmobranch heart. The dipnoi of lung fish, however, show an important variation. This order is only represented by a few scattered species, but these show the first primitive attempt of a water-living vertebrate to acquire the power of air breathing. They possess lungs, and the veins from the lungs open into a portion of the auricle, which is separated from the rest by an incomplete septum. The oxygenated blood coming from the lungs is thus separated in the heart from the oxygen-reduced blood that comes from the rest of the body.

7.3.1.5. Amphibian. The frog is a good example of this group because its heart has special structures and organization necessitated by air breathing. A special auricle has developed on the left where the aerated blood from the lungs is collected. The ventricle is not divided, but a complex system of valves in the bulbus arteriosus ensures that a large proportion of the aerated blood from the lungs passes to the head while a large proportion of the oxygen-poor venous blood from the rest of the body passes to the pulmonary arteries. The frog's heart is developed by the convolution and differentiation of a simple tubular structure. The mechanism of the frog's heart is one of the simplest examples of a double circulation, that is, of a structure by which the blood is alternately passed through the lungs to be oxygenated and then through the body to provide oxygen to the tissues. The actual mechanism for the separation of exhausted and oxygenated blood appears to be imperfect in the case of the amphibian heart.

7.3.1.6 Reptiles. The process of division of the circulation into two distinct circuits is carried to completion in the reptiles. In all reptiles there are two distinct auricles; in most genera the ventricle is only partly divided, but in the crocodile the ventricle is divided into two separate chambers and thus the pulmonary or lung circulation is completely separated from body circulation.

7.3.1.7. Birds and Mammals. The hearts of birds and mammals have evolved into more complex and powerful pumps than those of the cold-blooded vertebrates. The heart consists of four chambers, two auricles, and two ventricles. The sinus venosus of earlier more primitive forms is reduced to a small but very important piece of tissue known as the heart's pacemaker. The left auricle, which receives the blood from the lungs, is equivalent to the left half of the fish's auricle. The two ventricles have been derived by division of the single ventricle seen in more primitive heart forms, while the bulbus arteriosus of these earlier forms has disappeared. In birds' and mammals' hearts the venous blood returning from the body is collected in the right auricle and passes to the right ventricle which pumps it through the lungs. The lung veins return the blood to the left auricle and the left ventricle pumps the blood to the body tissue. The structure of the human heart is shown in Figure 7.6.

7.3.2. Capacity to Direct Energy

The data on the structures and organization of cardiovascular systems verify the fact that each system has its own characteristic capacity to pump blood to the cells of the body based on its specific structure and organization. An animal's cardiovascular system's capacity to direct energy results from the structure and organization of its heart. Quantification of an animal cardiovascular system's capacity to direct energy is determined principally by the structure and organization of the heart. Insight into the structure and organization of the vertebrate heart is obtained by a comparative analysis of its development.

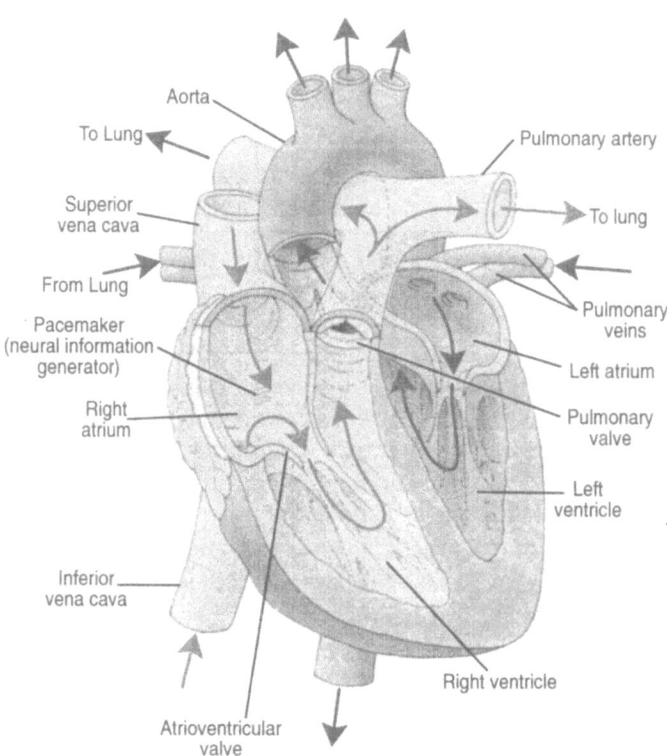

Figure 7.6. Structure of the human heart.

7.3.2.1. Development of the Vertebrate Heart. A comparison of the circulation in the different classes of vertebrates reveals systems of increasing complexity and efficiency. The hearts of the fishes (vertebrates) are more efficient and complex than the hearts of the higher mollusks (invertebrates). The emergence of land animals from water animals involved the substitution of lungs for gills, and necessitated an increase in circulator efficiency. The heart of a frog, for example, is larger, more complex, and much more efficient than the heart of a fish of the same size. Vertebrates have more powerful hearts to maintain a constant body temperature above the temperature of their environment. Higher body temperatures place greater demands on the heart because the rate of oxygen consumption is far higher in warm-blooded than in cold-blooded animals. In fact, the circulatory mechanisms of the birds and mammals are far more powerful and efficient than those of any invertebrates or cold-blooded vertebrates.

The hearts of warm-blooded animals, such as mice, have to work at a far greater pace than the hearts of cold-blooded animals such as frogs. For example, a frog's heart beats about 20 times per minute, whereas a mouse's heart beats from 500 to 1,000 times per minute. These high speeds necessitate a very careful timing of the contractions of the different portions of the heart. This is achieved by the specialization of certain tissues which conduct the impulses from chamber to chamber. The pacemaker of the mammalian heart generates an impulse that spreads all over the two auricles at a rate of about 40 inches per second, which is ten times the rate of the conduction in the frog's heart. The auricles are firmly attached to the ventricles, but there is only a small bridge of tissue that carries the wave of excitation from the auricles to the ventricle. This bridge is called the Bundle of His and consists of specialized muscle fibers called Purkinje fibers. The exact structure of the Bundle of His is well known in mammals. It is a well-defined bundle that first splits into two divisions which pass to the two ventricles and then branches out into a network of fibers covering the interior of the ventricles. The start of the Purkinje system is a mass of small muscle cells called the auriculo-ventricular node. Here the wave of excitation is slowed for a fraction of a second (in man for about one-twentieth of a second), and then the contraction wave is carried rapidly down the Purkinje system so that it reaches all portions of the ventricle at almost the same instant.

A heart's information transmission structure allows sufficient time for the blood to move from auricle to ventricle when the auricle contracts and at the same time ensures that the whole ventricle contracts simultaneously. The high efficiency of the mammalian heart depends very largely on this peculiar timing mechanism. The remarkable adaptability of the heart as a pump is illustrated by the fact that all mammalian hearts are built on the same structural plan, but the heart of a small mouse weighs about 0.15 g whereas the hearts of the largest whales weigh about 200 kg. Moreover, there are huge variations in speed—the heart of the mouse can beat 1,000 times a minute whereas the heart of the horse beats only 30 times per minute. These examples show the extraordinary range of size and speed over which the mammalian heart is capable of functioning efficiently.

7.3.2.2. Capacity of the Vertebrate Heart for Work. An outstanding feature of the heart is its capacity to perform continuous work, that is, to direct energy. It beats continuously during the life of the body. In the case of the human heart, it contracts 100,000 times per day and the left ventricle daily expels some 10,000 liters of blood working against a resistance of 120 mmHg. Another remarkable feature is the reserve power of the heart. The work that the heart does during bodily rest is only a fraction of the work it can do. The power of the body to perform long continued muscular exercise is limited by the amount of energy that can be supplied to the muscles, and this again depends on the quantity of blood that the heart can expel per minute. The heart of a human athlete during vigorous exercise can expel three times as much blood per minute as it expels during bodily rest. To perform this work the heart itself requires about one quarter of the blood it expels. The heart's oxygen consumption during such vigorous exercise is nearly equal to the oxygen usage of the whole body when at rest. Animals capable of long muscle endurance behaviors, such as the dog, the horse, the hare, and man are characterized by large hearts with great reserve powers. On the other hand, animals which never perform continuous exertion have relatively small hearts with low reserve powers. The rabbit and hare provide a striking contrast. The rabbit's heart weighs 2.7 parts per 1,000 parts of body weight, whereas the hare's heart weighs 7.5 parts per 1,000 parts of body weight. The rabbit's pulse rate at rest is 205, which is about two-thirds of the maximum rate at which it can work, whereas the hare's pulse rate at rest is only 64, which is less than one quarter of its maximum rate.

Since the maximum output of the heart is much greater than its normal output, it requires controlling mechanisms to adjust its work to the bodily needs. This adjustment is partly effected by variations in the venous filling of the heart— the heart muscle is organized so that the more complete the filling the more powerful the contraction. In addition the heart is controlled by two nerves: the vagus nerve, which reduces the heart's frequency, and the sympathetic nerve, which augments the frequency and the force of the beat. Augmenter and depressor nerves of this type are present in most invertebrate hearts and in all vertebrate hearts, but their activity is greatest in athletic animals. For example, cutting the vagus nerves increases the frequency of the hare's heart from 64 to 264 beats per minute, whereas cutting the vagus nerves in the rabbit only increases the frequency from 205 to 321 beats per minute.

Although the mammalian heart is a very efficient pump, there is a limit to the amount of work that can be done per minute by a given weight of muscle. The actual limiting fact is probably the maximum amount of blood, and consequently the maximum amount of oxygen, that can be supplied to the heart muscle per minute by the arteries of the heart (the coronary arteries). The oxygen required per unit weight by warm-blooded animals varies inversely as the cube root of their body weight, and hence small animals require much more oxygen per unit weight than do large animals. For example, the ox needs about 3.55 ml of oxygen per minute per kilogram of body weight, whereas the mouse needs 17 times this quantity.

It is a striking fact that a very large number of species of birds and mammals weigh between 10 and 30 grams but very few weigh less than 10 grams. A probable reason for this fact is that the oxygen requirements per unit weight in the case of animals of 10 grams is so great that the heart has to do work at a rate that approaches the maximum possible capacity of heart muscle.

7.3.2.3. Mammalian Hearts. In mammals the heart has become entirely divided into two halves, right and left, which have no direct communication with one another. The right auricle receives the venous blood from all parts of the body. From the right auricle, the blood passes to the right ventricle, and from there it is forced into the lungs along the pulmonary artery. In the lungs it takes up oxygen and becomes arterial, and then is returned by the pulmonary veins to the left auricle and to the ventricle. The left ventricle forces the blood into the arteries which carry it to all parts of the body.

There are two circulations: (1) the pulmonary, from the right side of the heart to the pulmonary artery and the capillaries of the lungs and to the left side of the heart by the pulmonary veins, and (2) the systemic, from the left side of the heart, by the aorta, to the arteries and capillaries of the body tissue and organs, then by the veins to the right side of the heart. The muscular walls of the right ventricle are much thinner than those of the left ventricle. This is so because the energy required of the left ventricle must exceed that of the right ventricle, because resistance in the systemic circuit exceeds that in the pulmonary circuit.

The heart becomes filled with venous blood during its relaxation, or diastole, and forces the blood into the arteries during its contraction, or systole. The large arteries have less capacity than the corresponding veins, and their walls are extensible and elastic. The small arteries and arterioles are muscular tubes and can vary considerably in diameter. The arterioles open into the capillaries, which are so numerous that, when taken together, can hold a large amount of blood. The skeletal muscles and the muscular walls of the viscera at each contraction compress the blood within the viscera and thereby influence the circulation. The whole muscular system is regarded as an accessory pump to the vascular system. The veins are larger in cross-section than the corresponding arteries and have tough and inflexible walls. The veins are not, as a rule, distended with blood to their full potential capacity. The latter is so great that all the blood of the body can collect within the veins.

7.3.2.4. Quantification of the Capacity to Direct Energy. The structures and organizations of the various cardiovascular systems and hearts described above provide these systems with their own behavioral characteristics. Specifically, the structure and organization of a particular cardiovascular system provides it with its capacity to direct energy—a fundamental characteristic of a living system. This capacity to direct energy is a direct function of the amount of work the heart can perform. The capacity of a heart to direct energy can be calculated using Equation 7.1.

The energy of the contraction of the heart is expended (a) in forcing a certain amount of blood against a certain resistance presented by the arterial pressure, and (b) in imparting a certain velocity to the blood. The work done by each ventricle can be calculated from the formula:

$$W = Mr + (Mv^2/2g) \tag{7.1}$$

where W is the work done, M is the mass of blood expelled at each beat, r is the mean arterial pressure in millimeters of mercury, v is the velocity at the root of the aorta, and g is the factor of acceleration.

The work of the right ventricle is approximately one-fifth of that of the left. The work of both ventricles in the human heart at rest is about 100 gram-meters per beat, which is equivalent to about 10,000 kilogram-meters in 24 hours. During very strenuous muscular exercise when the output is greatly increased, the work of the heart per beat is about 400 gram-meters, or 80,000 kilogram-meters in 24 hours. This rate of work could probably be maintained for not more than a few minutes.

7.3.3. Available Energy

The energy required for cardiac contractions is derived from the oxidation of glycogen within the heart itself. The maximum efficiency of the heart is of the same order as that for skeletal muscle, namely 20 to 28%. Efficiency is determined by comparing the work energy with the chemical energy input to the muscle. The remainder of the chemical energy used by the muscle is in the form of heat energy.

7.3.4. Neural Information Generation and Distribution

The concept of autonomous cardiovascular behavior depicted in Figure 7.1 is based on information being generated in the heart. Validation of this concept and quantification of autonomous behaviors require an understanding of the generation and distribution of information associated with these behaviors. The extant data on the information associated with cardiovascular behaviors were used to validate this concept.

The cause of the heartbeat has long been a subject of inquiry. In 1757, H. Allen was the first to show that the activity of the heart is not dependent on its connection with the nervous system. The heart is controlled and influenced by the nervous system, but this control is not essential for life. The excised heart of a frog continues to beat rhythmically for days, provided that it is supplied with oxygen and prevented from drying. In the case of the warm-blooded animal, the heart is similarly capable of continuing its rhythmic contractions for some time after excision.

7.3.4.1. The Amphibian and Reptilian Heart. Experiments have shown that every strip of the frog's heart muscle is capable of rhythmic action, whether it contains nerve cells or not. In the developing chick, the heart contracts within 24 to 28 hours of incubation, while the nerves do not grow into the heart before the sixth day. These data demonstrate that the inherent power of rhythmic contraction belongs to the cardiac muscle itself (the "myogenic" theory of the heartbeat). Furthermore, it belongs to every part of the heart, but there is a descending scale of this automatic power, from the sinus where it is highest to the lower parts of the ventricle where it is slight (Gaskell). The normal sequence of contraction of the four parts is determined by the natural rhythm of these parts, but in the whole heart it is impossible for the ventricle to contract at its own rhythm because before it is ready to beat again, after a preceding contraction, it receives an impulse from the auricle. In the same way, the auricle never beats at its own rhythm; it is always subordinated to the faster rate of impulses coming from the sinus. If, however, the ventricle is electrically stimulated at a rate slightly faster than the beat of the sinus, the normal sequence of contraction becomes reversed, the ventricle now contracting first and the sinus last. These data confirm the myogenic theory of the heartbeat.

7.3.4.2. The Mammalian Heart. Although the mammalian heart has no sinus venosus, its contractions are as regular and as independent of the nervous system as those of the lower vertebrates. Both in the heart *in situ* and in the excised heart, the two auricles contract together, and after a short interval follows the contraction of the two ventricles.

In the mammalian heart, within the region where the superior vena cava opens into the right auricle, there lies a club-shaped formation known as the *sinoauricular* or the S-A node; functionally, it is identical to the sinus of the amphibian and reptilian heart. The node is composed of slender fusiform cells with little striation. The sinus serves as the center in which the information for the cardiac contraction originates. The S-A node is, however, not the only place in which the information for the heart muscle contraction can originate. The seat of these rhythmically recurring information quanta may shift to some other portion of the heart since, as in the frog's heart, the function of information generation is potentially present in every part of the heart, and the S-A node governs the rate of the whole heart only by virtue of its faster rate of information generation. On this account the S-A node is described as the pacemaker of the heart. Other centers in which information sometimes originates, and which, in certain cases, may gain mastery over the whole heart, are known as *ectopic* centers.

It has been shown that a very distinctive system of muscle fibers lies enclosed within its own sheath beneath the endocardium. This system is known as the *conductive system* of the heart. It begins as a few strands of fibers in the region of the coronary sinus; these strands converge in a thickening which is known as the *auriculo-ventricular* or the A-V node; it is composed of the same type of cells as the S-A node. The two nodes are not connected with each other but are divided by the ordinary contractile elements of the auricle. From the A-V node, a thin bundle of tissue passes through the A-V septum toward the ventricle. This bundle was first described by His, and is known as the Bundle of His. On penetrating the intra-ventricular septum, the bundle divides into two branches, which pass to the right and left ventricle, respectively. The two branches in turn divide and subdivide, forming an extensive network on the inner surface of the ventricles.

Various parts of the conductive system may become centers of ectopic rhythms. If the S-A node is destroyed or injured, or if the irritability of the A-V node is increased, the latter assumes the role of initiating the heartbeat. Many nervous and other influences may modify the relative irritability of the two nodes, and a shift of the pacemaker from one to the other may occur temporarily even under normal conditions. Destruction of the His bundle is equivalent to a complete functional separation of the ventricles from the auricles. The ventricles develop a rhythm of their own, the *ido-ventricular* rhythm. Both ventricles, however, continue to contract together, the center for their activity being localized in the higher part of the bundle.

Thus, every part of the heart may serve as a center of origin of an information quantum. The natural rate of information generation of these centers is in the following descending order: the S-A node, the A-V node, the bundle, its branches, and muscle. If several centers are active simultaneously, the rate of the heart as a whole will be dominated by that center with the highest information generation frequency.

7.3.4.3. The Normal Heart Rate. The normal heart rate varies between individuals and between different species of animals. For a man, it may be 68 to 76 beats per minute, and for a woman 74 to 80; but the normal rate for some individuals may be much lower (50) or much higher (90). Small animals, as a rule, have a higher heart rate than large animals (e.g., elephant 25 to 28, horse and ox 36 to 50, sheep 60 to 80, dog 100 to 120, rabbit 150 to 180, mouse 700; small birds like the canary have an extremely high rate of 1,000 beats per minute). Usually in man, under normal conditions, the heart rate declines with age. At birth it is about 140, it is 100 to 110 at the age of 5; in late childhood it is about 90, and in the adult about 70. In old age it accelerates slightly to about 80.

7.3.4.4. Information Distribution. Information that originates in the S-A node spreads along the ordinary muscular tissue to every part of the auricles. The rate of conduction in the auricle ranges from 600 to 1,200 mm per second. In its fan-like spread, the information reaches the A-V node. There is no indication of a preferential path of conduction between the two nodes, so that the information travels at the same rate over the auricle and reaches the A-V node about 0.03 seconds after origination. In the A-V node, the rate of conduction is considerably slower, the information passing the node at about 150-200 mm. per second. Once it has passed the node, it travels rapidly down the bundle and its branches at the rate of about 5,000 mm per second.

The slow conduction in the A-V node and the rapid conduction in the bundle tissue ensure two important features of cardiac activity. On account of the former, auricular contraction is given time to end before the onset of the ventricular contraction, and on account of the latter, the information arrives at every part of the ventricle at approximately the same time. Thus the whole ventricular muscle contracts approximately at the same time, which is a condition necessary for the development of a high pressure in the ventricular cavities.

7.3.5. Cardiovascular Behavior

The last element of the behavioral concept of the cardiovascular system is cardiovascular behavior itself, and the relationship among capacities to direct energy, available energy, information, and behavior. The primary behavior of the cardiovascular system is the dynamics of pumping blood through the system.

For any given cardiovascular system, a quantum of information generated within the heart is distributed to all the contractile tissue in the heart which causes a maximal contraction of the heart. This maximal contraction behavior of the heart in response to a quantum information input is known as the "All or None Law." As explained in Chapter 4 the contractile behavior of motor units of skeletal muscle also follows this All or None Law when it receives a quantum of information.

In living systems behavior terms, the behavior of pumping blood through the cardiovascular system is a function of the specific capacity to direct energy characteristic of the cardiovascular system, the energy available to the system, and behavioral information inputs to the heart to cause its contraction. Each of these parameters can be measured or quantified. The data presented below verify this concept of behavior.

The behavior of the cardiovascular system for one heartbeat is as follows. One quantum of behavioral information is generated in the heart and distributed, in a specific way depending on the characteristics of the heart, to the contractile tissue in the heart. This tissue contracts, causing blood to be passed through the heart and into the vascular system. The total energy in the contractile behavior is the sum of the mechanical work done by the heart to pump the blood and the heat energy that is generated during the heart contraction. This behavior is very similar functionally to that of a skeletal motor unit, with a few differences: (1) the quantum of information for the cardiovascular system is generated in the heart whereas the quantum of information for the motor unit is generated external to the motor unit and is transmitted to the motor unit via a motoneuron, and (2) the work of the skeletal motor unit is in the form of mechanical motion, whereas the work in the heart behavior is in the form of moving a mass of blood against an arterial pressure. In both the heart and the motor unit contractile tissue, heat energy is generated during each contraction. The behavior of one heartbeat can be quantified by modifying the equations shown in Chapter 5 and combining those equations with Eq. 7.1 of this chapter.

7.3.5.1. The Refractory Period. Continual heartbeat is more complex than just a single beat due to additional phenomena, such as the refractory period. Heart behavior undergoes rhythmical variations which are determined by characteristics of the heart itself. Like all excitable tissue, the heart has a *refractory period*, which is a period of loss of excitability following each information quantum that causes a contractile response. In the heart this period is extremely prolonged. It lasts as long as the contraction of the heart. If a second quantum of information is applied to the heart muscle during the contraction which is caused in response to the first information quantum, it has no effect. The period of complete loss of excitability is followed by gradual recovery, after which there is a phase of supernormal excitability before it returns to normal.

The length of the refractory period depends on the strength of the contraction of the heart. Certain drugs and physiological conditions which strengthen the heartbeat also increase its refractory period. Of greater significance, however, is the relation of the refractory period to the rate of the heartbeat. Lewis (1911) found that the duration of the refractory period of mammalian auricular muscle contracting 100, 130, and 250 times per minute was 0.2, 0.15, and 0.01 second, respectively.

7.3.5.2. Contractility. The strength of contraction of the heart is highly dependent on the condition of the cardiac muscle. Any interference with the energy available to the heart—the nutrition of the heart, inadequacy of the oxygen supply, or accumulation of the products of metabolism such as carbonic acid or lactic acid—will lead to a weakening of the contractile response, and even to a complete loss of contractility.

There is also a physiological factor which modifies the strength of contraction. Briefly, other conditions being equal, the greater the filling of the heart during diastole, the stronger the following systole. This dependence of strength of contraction is so pronounced and is of such importance for the whole circulation that Starling named it the "Law of the Heart." It is not a feature peculiar to the cardiac muscle alone, but belongs to all contractile tissues, including skeletal muscle and smooth muscle. In the heart it is of vital importance. The practical significance is obvious: it enables the heart to eject the amount of blood it receives during diastole, whether small or large, and thus enables the heart to adapt its beat to large variations in blood flow without changing its rate. A dog's heart weighing 50 grams can put out 100 or 3,000 ml of blood per minute without changes of the heart rate. This remarkable adaptation is due to the fact that the larger the output, the more the heart will be filled during each period of relaxation (diastole), and hence its contraction (systole) will be stronger, so that the heart will empty itself of the extra amount of blood.

7.3.5.3. Sequence of Events in the Cardiac Cycle. The time relation between the different events occurring during a cardiac cycle can be determined by measuring pressure changes in the different cavities of the heart and in the aorta. These pressure changes, along with their timing, can be used to determine the blood velocities and acceleration factors for use in Eq. 7.1.

The pressure changes in the heart have been described by Starling as follows: The cardiac cycle begins with the contraction of the auricles, which may or may not cause a slight rise of pressure in the ventricles. As the auricular contraction dies away, the ventricular contraction begins. This causes a very rapid rise of pressure. Almost immediately after the beginning of the rise, the auriculo-ventricular valve closes. The pressure then rises rapidly in the ventricular cavity. During this period, the contraction of the ventricular muscle raises the pressure within the ventricles without causing any change in its contents or in the length of the muscle fibers. Soon the pressure exceeds that in the aorta, the aortic valves open, and the aortic pressure rises thereafter with the ventricular pressure. The ejection of blood is at first rapid so that the pressure in the ventricles continues to rise. As the heart gets smaller, the amount of blood ejected into the aorta becomes less than that flowing out through the peripheral branches, so the pressure begins to fall in the aorta and ventricle even though the outflow of blood is still going on.

The pressure in the ventricles then continues to fall more slowly until it reaches zero; it remains at or near zero during the greater part of diastole. With a big inflow there may be a slight rise toward the end of diastole, which may be accentuated by the auricular contraction.

The duration of the separate phases of the heartbeat depends on the rate of the beat. In all cases of change of heart rate, the period of diastole is affected relatively more than the period of systole. Because there are no valves between the auricles and the large veins, changes of pressure within the auricle are transmitted along the veins.

In every case, an auricular pressure tracing exhibits the following features: (1) the first positive wave, which occurs during auricular systole; (2) the second positive wave, which is due to the sharp closure of the auriculo-ventricular valves; (3) the third positive wave, which is due to the filling of the auricles while the auriculo-ventricular valves are closed; and (4) a negative wave, which is due to the rapid emptying of the auricle after the opening of the auriculo-ventricular valves. The main function of the auricle is not to propel blood into the ventricle by its contraction, but to serve together with the big veins as a reservoir for the blood which flows in from the body but which cannot enter the ventricle while the latter is in a state of contraction.

7.3.5.4. The Nervous Regulation of the Heartbeat. In the vertebrates, the heart is supplied with two sets of nerve fibers: those which pass from the central nervous system in the vagus nerve, and those which pass in the sympathetic nerves. The cardiac fibers of the vagus terminate around nerve cells situated in the heart itself (preganglionic fibers); the ganglionic cells serve as relays from which new fibers (postganglionic fibers) emerge and run directly to the cardiac muscle and to the S-A node. The sympathetic fibers leave the spinal cord by the anterior roots mainly of the second and third dorsal nerves, run in the white *remi communicants* to the stellate ganglia where they end. From the stellate ganglia, postganglionic fibers begin, which go to the various parts of the cardiac muscle.

From a behavioral information point of view, two new information generators have been introduced which can cause modified behaviors of the cardiovascular system. This new information, when combined with the information generated in the heart, can cause the heartbeat to slow and to accelerate.

7.3.5.5. The Vagus Nerve. In 1845, the Weber brothers made the important discovery that stimulation of the vagus nerve retards or even arrests the heartbeat. The cardio-inhibitory nerves have since then been found in all classes of vertebrates and in many invertebrates. During a stimulation of the vagi, the heartbeat is retarded or stops altogether, and as a result arterial blood pressure falls. If the stimulation of the vagus is prolonged, the heart often begins to beat again with a slow rhythm; this beat is confined to the ventricles only. The ventricle is actually beating at its own rhythm in response to information originating within itself.

Vagus stimulation affects all the four fundamental properties of the heart muscle. It (1) inhibits the S-A node, depressing the rhythmicity; (2) affects the conductive system, retarding the propagation of the information quantum in the auricular muscle and in the A-V bundle; (3) affects the muscle proper, decreasing its contractility, whereby each beat becomes weaker and the refractory period of the heart is shortened, and (4) diminishes the excitability of the heart.

If both vagi are cut, the heart immediately begins to beat faster, showing that under normal conditions a continuous stream of information quanta passes down the cardio inhibitory nerves, which do not allow the heart to beat at its full independent rate.

Control of the normal tone of the vagi is a function of blood pressure. Marey was the first to show that, other conditions being equal, the vagus tone increases with an increase in the blood pressure. This effect is probably due to stimulation by *high blood pressure of special sensory endings in the aorta, in the ventricles,* and in some of the blood vessels going to the brain. The information generated by these sensors retards the heart by stimulating the vagus center. Changes in the composition of the blood and various drugs may also affect the vagus center. Asphyxia and the action of morphine retard the heart, but only if the vagi are intact. Reflexes from various sensory nerves may stimulate or inhibit the vagus center. For instance, inflation of the lungs diminishes vagus tone (the Hering Breuer reflex); an increase in the output of the heart has the same effect (Bainbridge reflex). Stimulation of the respiratory passages, as in the case of inhalation of an irritant volatile substance, retards the heart. There are also stimuli which may excite the peripheral nerve endings of the vagi in the heart itself—for instance bile salts, which enter the general circulation in the case of jaundice.

7.3.5.6. The Sympathetic Nerve. Stimulation of the sympathetic cardiac nerve produces effects which are the opposite of the vagus stimulation. It increases the rate by raising the rhythmicity of the heart, augments the contractions, increases the rate of conduction of the impulse, and raises the excitability of the heart. On account of the increased strength of contraction, the refractory period becomes somewhat more prolonged than that normally associated with the given heart rate. The sympathetic nerves are less easily tired than the vagus fibers and have a longer after-effect. In most animals the inhibitory and the accelerator fibers become mixed in the cardiac nerves, so that if these are stimulated, a double effect is produced on the heart. During the period of stimulation the vagus effect predominates, but after the end of stimulation the sympathetic effect becomes apparent, and the heart accelerates and the beat increases in strength.

Stimulation of either set of fibers before they are mixed together produces effects typical of one or the other only. In the dog, whose normal heart rate is about 100 beats per minute, stimulation of the accelerator nerves may increase it to 260 beats, and at the same time the strength of each contraction will be increased.

7.3.5.7. The Suprarenal Gland. The suprarenal (or adrenal) gland produces a chemical substance which can stimulate all the sympathetic nerve endings. The substance is adrenaline or epinephrine, which is a hormone, that is, chemical information. Epinephrine is active in very small amounts—concentrations of one in one hundred million produce a strong exciting effect. Under normal conditions, however, the quantities entering the blood are probably too small to have any physiological effect. The liberation of epinephrine is under the influence of the splanchnic nerves—inhibiting these nerves diminishes the secretion while stimulation greatly increases it. Certain drugs, asphyxia, and various emotions such as fear or anger lead to an excessive production of epinephrine. Epinephrine that reaches the heart causes the contractions to become considerably faster and extremely energetic, and renders the heart capable of coping with a greater strain (either in the form of arterial resistance or increased venous inflow).

7.3.6. Quantification of Cardiovascular Behavior

The data presented above are those necessary and sufficient to validate the concept of the autonomous cardiovascular behaviors depicted in Figure 7.1. The fundamental parameters relating to these behaviors are also firmly established and can be quantified. The relationships among these parameters are also firmly established, can be quantified, and have the highest levels of confidence.

7.4. The Respiratory System

Breathing is part of respiration, which is a process common to all forms of animal life. Respiration is essentially the transport of oxygen from the air or water to the place where the oxygen is used up by the body, and the transport of carbon dioxide from the place where it is produced to the external environment. Many animals live in water but even for them the ultimate source of oxygen is the atmosphere; from it the water acquires new supplies of oxygen as the animals that inhabit it use up the gas. The oxygen in water is for the most part in solution, not in bubbles; in the sea the constant breaking of the waves oxygenates the surface layers of the water.

In the most primitive forms of life, such as the amoeba, respiration is very simple. The amoeba lives in water. Oxygen from the water soaks into the body of this animalcule, where it is always being used up. Because oxygen is always being used, the concentration of oxygen inside the amoeba is always less than the concentration of oxygen in the water outside. The oxygen, therefore, by a simple process of diffusion, migrates from the place of higher to that of lower concentration, that is, from the water to the interior of the amoeba, so a constant stream of gas is maintained. Carbon dioxide produced in the amoeba diffuses out through its surface into the surrounding water, that is, from a high to a low concentration of carbon dioxide.

In the higher forms of life, the diffusion principle is no different from that of the amoeba. The systems for effecting respiration become more complicated, but the actual process is the same, namely, the diffusion of gas, oxygen or carbon dioxide, from the place of higher to the place of lower concentration. It is these more complex respiration systems that are discussed herein, in particular, animals with nervous systems.

Hypothetically, the basic concepts of breathing behaviors, described early in this chapter and depicted in Figure 7.3, would allow these behaviors to be quantified and that there is sufficient data to test and verify the concepts. The test performed to verify these hypotheses includes the following four steps. First, the structure of the respiratory system of specific animals is identified and specified—it is structure and organization that determines a respiratory system's capacity to direct energy. Second, the information generation and distribution system which causes the respiratory behavior is identified. Third, the energy available for the behaviors is identified and specified. Lastly, the energy in the behavior is identified and specified.

7.4.1. Respiratory Structures

Quantification of respiratory behavior depends on the structure and characteristics of the respiratory system. There are a number of major types of respiratory structures, ranging from the very simple to the very complex. These include arthropod, mollusk, fish, amphibian, bird, and mammalian structures. A short review of these structures is given below.

7.4.1.1. Arthropod. Arthropods that breathe by tracheae, such as insects, have respiratory structures very different from other animals. In them, respiratory air is piped all over the interior body to, or almost to, the actual functioning cells. There is no intermediary circulating fluid. The whole tissue of the insects is permeated by an elaborate system of tubes, the tracheae, with walls stiff enough to prevent their collapsing. The tracheal tubes are often extremely small in cross-section. This system has limitations. The rate at which gases can diffuse along very fine tubes is slow, and sufficient oxygen can only penetrate a short distance. No portion of the insect can be far removed from the external air, and for that reason all insects are small; the largest one is the dragonfly, which has a relatively long but extremely thin body.

The tracheal tubes usually open at the surface of the body through a number of spiracles. A sphincter is associated with each spiracle and is opened only at intervals as the need for oxygen arises. The sphincter is composed of muscle tissue and the motoneurons associated with this muscle tissue. In the smaller and less active insects, oxygen passes from spiracles to tissues wholly by diffusion. But in larger and more active forms (e.g., grasshoppers, bees, and flies), the tracheae are ventilated by alternate contraction and expansion of the body wall. The tracheae walls contain muscle tissue and their associated motoneurons which make these contractions possible. These respiratory movements occur particularly during activity, and in many insects they are made more effective by the expansion of the tracheae at various points to form large, thin-walled air sacs. These are ventilated mechanically, but the supply of oxygen through the branches running from them is still dependent on gaseous diffusion. This mechanical ventilation of the air sacs is made possible by muscle tissue and their associated motoneurons which provide muscle contraction. The increased opening of the spiracles and the increased mechanical ventilation during activity are brought about by the accumulating carbon dioxide causing the generation of information in the respiratory centers in the nervous system.

Arthropods that breathe by gill have more complex structures than those that breathe by tracheae. A circulation of fluid, which differs little from seawater, is maintained throughout the animal. In the gill, this fluid is separated from the seawater only by the wall of the vessel in which it is flowing. That wall is no thicker than the body of the amoeba, so the gases (oxygen and carbon dioxide) easily diffuse into and out of the circulating fluid. The amount of gas which can be dealt with depends principally on the extent of the surface of circulating fluid that can be exposed to the water at any one time. The gills of some creatures, such as the lobster, have feathery forms which provide the maximal surface area. Indeed, the surface is so great that the water around the gill would be completely depleted of oxygen were there not a special mechanism for ensuring a continuous supply of fresh oxygen-charged water over the surface of the gill feathers.

The gill system of the lobster exhibits the principles on which the respiratory systems of almost all the higher animals are based, namely, the exposure of a large surface of fluid which circulates in the animal (the blood) to a corresponding large surface of either air or water which is constantly being replenished. The oxygen-containing medium and the circulating blood are not in actual contact but are separated from one another by a membrane through which the oxygen (and carbon dioxide) must diffuse.

The respiration circulation system for arthropods with gills is part of the cardiovascular system, as described in the earlier cardiovascular section. There is no separate pumping system for breathing. At this level of organization, there is no separate neural/muscle system for respiration behavior.

7.4.1.2. Mollusk. The respiration structures of higher mollusks are more complex than those of the arthropods. For example, in cephalopods (e.g., octopus), the circulatory structure includes accessory hearts in addition to the main heart. These accessory hearts pump blood through the gills where the blood is oxygenated. From a behavioral point of view, the behavior of these hearts can be treated as similar to the hearts of the cardiovascular systems.

7.4.1.3. Fish. The respiration structures of almost all species of fish include internal gills. These gills have a large surface area for gas exchange between blood and water. They are supported by five or six bony gill arches on either side of the fish between the mouth cavity and the protective opercular flaps. Each gill arch is lined with hundreds of leaf-shaped gill filaments arranged in two rows. These rows of gill filaments point toward the opercular opening, which is the direction of water flow, and the tips of the filaments of adjacent arches are interdigitated. The upper and lower flat surfaces of each gill filament have rows of evenly spaced folds, or lamellae, which greatly increase the gill surface area. The large surface area of the lamellae is the site of gas exchange. The result of this interdigitating network of gill filaments and lamellae is that practically all water that passes across the gills comes into close contact with the gas-exchange surfaces.

The delicate structure of the lamellae minimizes the path length for diffusion of gases between blood and water. Blood travels in blood vessels through the gill arches and the gill filaments, but in the lamellae the blood flows between the two surfaces of the lamellae as a sheet not much more than one red blood cell thick. The surfaces of the lamellae consist of highly flattened epithelial cells and practically no cytoplasm, so the water and the red blood cells are separated by little more than one or two micrometers.

The rate of diffusion is maximized by ventilating both the external and the internal surfaces of the lamellae. A constant flow of water moving over the gills maximizes the oxygen (O_2) concentration on the external surfaces. On the internal side, the concentration of O_2 is minimized as the circulation of blood sweeps the oxygen away as rapidly as it diffuses across the lamellae.

Most fish ventilate the external surfaces of their gills by means of a two-pump mechanism that maintains a unidirectional and constant flow of water over the gills. The closing and contracting of the mouth cavity act as a positive-pressure pump, pushing water over the gills. The opening and closing of the opercular flaps act as a negative-pressure pump, pulling water over the gills. Because these pumps are slightly out of phase with each other, they maintain an almost continual flow of water across the gills. Thus, moving water is always bathing the gill surfaces. The structure of these pumps includes motor units for closing and contracting the mouth cavity and for opening and closing the opercular flaps.

There are, however, a few species, such as the lung fishes, that have lungs which provide the capability of breathing air. The veins from the lungs open into a portion of the auricle of the heart, which is separated from the rest by an incomplete septum. The oxygenated blood coming from the lungs is thus separated in the heart from the reduced blood that comes from the rest of the body. The structures of lungs are treated in the more complex animals described below.

7.4.1.4. Amphibian. The respiration systems of amphibians are capable of extracting O_2 from both water and air. The principle of extracting oxygen from water is treated above. The principles of extracting oxygen from air are explained using an air-breathing amphibian, the newt, as an example. The lung of the newt is somewhat like a small balloon. Blood circulates in the wall of the lung, with a large amount being exposed in a close network of capillaries to the air in the lung. Air is intermittently forced in and out of the lung by swallowing movements on the part of the newt, the stalk of the lung (or trachea) being an outgrowth of the gullet. The swallowing movements are made possible by the muscle and their motoneuron structures associated with the swallowing behavior.

7.4.1.5. Birds. The respiration structures of birds include air sacs at several locations in their bodies and lungs. The air sacs connect with one another and with the lungs. Even though the air sacs as well as the lungs receive inhaled air, the air sacs are not gas-exchange surfaces.

The microscopic anatomy of the bird lung is unique among air-breathing vertebrates. The common lung design of other air-breathing vertebrates is a system of branching airways that get finer and finer until they end in clusters of microscopic air sacs where gas exchange takes place. Air flow in such a system is tidal—it enters and leaves by the same route. The bird lung, in contrast, does not have blind air sacs, but has tube-like parabronchi that permit unidirectional air flow through the lungs. Air capillaries off the parabronchi increase the surface for gas exchange. Another unusual feature of bird lungs is that they expand and contract relatively little during a breathing cycle.

Another difference between bird and mammals is that a single breath of air remains in the bird's gas-exchange system for two inhalation–exhalation cycles. On the first inhalation, the inhaled air goes primarily to the posterior air sacs. When the bird exhales, the air in the posterior air sac flows through the parabronchi of the lungs. Upon the next inhalation, the air in the parabronchi flows into the anterior air sac, as fresh air fills the posterior air sacs. Finally, the air in the anterior air sac leaves the bird on the exhalation after it entered. The two sets of air sacs act as bellows, maintaining a constant, unidirectional flow of air through the lungs. The structures of the lungs and air sacs include the motor units which cause contraction and expansion of the lungs and the air sacs.

7.4.1.5. Mammals. The respiratory structures of mammals were adequately described earlier in the section on the concepts of breathing.

7.4.2. Capacity to Direct Energy

The respiratory structures and organizations identified and described above for arthropods, mollusks, fishes, amphibians, birds, and mammals provide these animals with their unique capacities to direct the energies required for respiratory behaviors. Verification of the principles of a quantitative living systems science, as they apply to respiratory behaviors, requires the quantification of these capacities to direct energy for respiratory behaviors.

Quantification of capacities to direct energy for respiration behaviors was performed using the methods and procedures of the development models presented in prior chapters. A basic premise of these models is to treat first the most readily observable and common phenomenon, which is breathing. Breathing, which is caused by contractions of the muscular respiration pump structures, is explained above.

Respiration pumps were considered to be in two categories based on the fluid being pumped—one pumps liquids and the other pumps gases. In general, animals with gills have respiration systems that pump liquids and animals with tracheae and lungs have respiratory systems that pump gases. Although the characteristics of all fluid pumps are generally the same, the differences between liquids and gases influence the more detailed characteristics of these pumps. Within each of these categories of pumps, there are many different structures with a wide range of sizes. These differences in structures and sizes determine an individual pump's capacity to direct energy.

7.4.2.1. Liquid Pumps. Liquid respiration pumps are of two general types: those that produce a pulsed flow and those that produce a continuous flow of liquid. The general structure for pulsed flow is a chamber with valves to control the flow of liquids encased in muscle tissue which contracts and relaxes to change the volume of the chamber. These structures have the capacity to direct energy and are similar to those of the heart pumps described previously. The capacity to direct energy of liquid respiration pumps is expressed by Eq. 7.1 with an appropriate interpretation of the terms in this equation, as follows.

The energy of contraction of the liquid respiration pump is expended (a) in forcing a certain amount of liquid out of the pump against a certain resistance presented by the pressure of the liquid in the circulation system, and (b) in imparting a velocity to the liquid. The work done by the pump can be calculated from Eq. 7.1, where W is the work done, M is the mass of the liquid expelled at each muscle contraction, r is the mean pressure in the circulation system, v is the velocity at the output of the pump, and g is the factor of acceleration.

Continuous flow of liquid in respiration pumps is based on structures consisting of two pumps operating together. The respiration systems of fishes typify this flow and the pumps which cause this flow. The work performed by the two-pump continuous respiratory system can be calculated by the formula:

$$w = pV \tag{7.2}$$

where w is the sum of the work performed by the two pumps during each contraction of the pump muscles, p is the differential pressure between the high and low pressure sides of the gills, and V is the volume of fluid passing through the gills during the time period from one initiation of a muscle contraction to the next initiation. It is the structure of these pumps that give them the capacity to direct energy (perform the work) required for the liquid flow.

7.4.2.2. Gas Pumps. Gas respiration pumps, like liquid pumps, are of two types: those that supply a varying amount of gas over diffusion tissue and those that supply a constant amount of gas. Human breathing, like that of many other vertebrates and like many arthropods, is of the varying, or, tidal, type. Birds, on the other hand, have a constant flow of gas over the diffusion tissue of their lungs.

Both liquid and gas pumps are fluid pumps and have many common characteristics, the differences between the characteristics being a function of the density and compressibility of gases as compared to liquids. When these differences are taken into consideration, the equations for pulsed and continuous liquid pumps given above can be tailored for application to gas pumps.

7.4.3. Available Energy

Just as with other muscular activity, chemical energy must be available for conversion to heat and mechanical work by the muscles of the respiration fluid pumps. The mechanical energy required for these pumps can be obtained from the above equations and the characteristics of the specific pumps under consideration—on a per contraction basis. Chemical energy must be available for these contractions to occur. The amount of this energy, on a daily basis, can be calculated from the energy used in each contraction multiplied by the number of contractions per day. The chemical energy available for pump muscle contractions is in the form of glycogen. The equations for the conversion from glycogen to heat and mechanical motion are well developed and understood.

7.4.4. Neural Information Generation and Distribution

Unlike the human heart, which has a self-contained information generator, primary breathing information is generated in the central nervous system, specifically, in the medulla part of the brain stem. If the spinal cord is severed just below the medulla (the segment of the brain stem just above the spinal cord), the information distribution system between the medulla and the breathing muscles is cut and breathing stops. If the brain stem is cut just above the medulla, irregular information is transmitted to the breathing muscles and a crude breathing rate results.

The primary breathing information generator behaves as follows: Groups of neurons within the medulla increase their firing rates just before an inhalation begins. As more and more of these neurons fire, and fire faster and faster, the inspiratory (inhalation) muscles contract. Suddenly the neurons stop firing, the inspiratory muscles relax, and exhalation begins. Exhalation is usually a passive process that depends on the elastic recoil of the lung tissue. Breathing information generated in the medulla is passed down the nerves which connect the medulla with the muscles of respiration. The principal nerves in question are the phrenic (fourth cervical) nerves, which supply the diaphragm, and the intercostal nerves, which supply the intercostal muscles. If the mechanism consisted merely of this primary information transmitted to the motor nerves which it controls, respiration would be a series of gasps at slow intervals. However, there are other information generators, each of which modifies the natural gasping information rate (rhythm). One such information generator in the brain imparts a smoothness to the gasping rhythm and a rate which gives respiration a more normal character. Other information generators in the brain area above the medulla can modify the ventilatory rhythm. The information rate, or rhythm, is modified to accommodate speech, ingestion of food, coughing, and emotional states.

As respiratory demands increase, the activities of the inspiratory neurons also increase, thus contributing to a greater depth of inspiration. There is an override reflex, however, that prevents the ventilatory muscles from overdistending and damaging the lung tissue. This reflex is called the Hering–Breuer reflex for the two physiologists who discovered it. It begins with stretch receptors in the lung tissue. When stretched, these receptors generate information which is transported by the vagus nerve that inhibits the inspiratory neurons.

7.4.4.1. Matching Ventilation to Metabolic Needs. When the metabolic rate changes the partial pressure of oxygen (pO_2) and the partial pressure of carbon dioxide (pCO_2) in the blood, the respiratory information changes so as to return these values to normal levels. It is reasonable to expect, therefore, that the blood pO_2 or pCO_2, or both, should provide information to the respiratory information centers in the brain.

Humans and other mammals are very sensitive to increases in the pCO_2 of the blood, whether it is influenced by energy demands or by the composition of the air breathed. Re-breathing a small volume of air so that the pCO_2 of that reservoir gradually rises, causes breathing to become deeper and more rapid. The reaction is the same even if pure O_2 is constantly released into the reservoir to keep its pO_2 constant.

It seems reasonable that CO_2 rather than O_2 should be the dominant feedback information for ventilation since animals have evolved respiratory systems and hemoglobin properties that work to keep the blood leaving the gas-exchange surfaces fully saturated with O_2 over a broad range of alveolar pO_2's and metabolic rates. Normal fluctuations in metabolic rate and ventilation have very little effect on the maximum amount of O_2 carried by the blood. In contrast, small changes in metabolism and alveolar pCO_2 do influence the concentration of CO_2 in the blood. Consequently, changes in blood pCO_2 are a much finer index of energy demands and respiratory performance than the O_2 content of the blood.

The major site of CO_2 sensitivity and information generation is an area on the ventral surface of the medulla, not far from the groups of neurons that generate the primary ventilatory information. Sensitivity to the O_2 concentration of the blood resides in the nodes of tissue on the large blood vessels leaving the heart, the aorta and the carotid arteries. These carotid and aortic bodies receive enormous supplies of blood, and they contain chemosensory nerve endings. If the blood supply to these structures decreases, or if the pO_2 of the blood falls dramatically, then the chemosensors are activated and send information to the primary respiratory information generator.

7.4.4.2. Other Information Generators. Sensors at the surface of the body can also generate information that is transported to the brain where it causes a change in the primary respiration information generator. For example, strong ammonia placed beneath the nose causes information to be generated which travels down the fifth cranial nerve and influences respiration. The sensory nerves from almost any part of the skin, too, can influence respiration, as when cold water is suddenly dashed on the surface of the body.

7.4.5. Respiratory Behavior

The primary respiratory behavior is a direct function of the behavior of respiratory pumps. As shown in Figure 7.2 a respiratory system's behavior is a function of its capacity to direct energy, available energy, and information.

For any given respiratory system, the behavioral concept is that respiratory system behavior can be quantified by knowing the system's fundamental parameters and their values. That is, we must know the system's capacity to direct energy, its available energy, and its information-generating characteristics. The relationships among the parameters of behavior and the behavior itself are described below for various classes of animals.

7.4.5.1. Arthropod. Arthropods have simple respiratory systems whose behavior consists only of the opening and closing of the sphincters of spiracles. These behaviors are the results of the contraction of sphincter muscles caused by inputs from information generators. If the sphincter is normally closed, muscle contractions are necessary to open the sphincter and to keep it open. Conversely, if the sphincter is normally open, then muscle contractions are necessary to close the sphincter and to keep it closed. In either case, a quantum of information will cause a contraction of the sphincter muscle tissue associated with the neuron which conducts the quantum of information. The amount of energy used in the contraction is determined by the structure of the sphincter muscle and can be measured or calculated. If the sphincter consists of multiple motor units, then a quantum of information is transmitted to each motor unit involved in the sphincter contraction. A continuous flow of information quanta is necessary to keep the sphincter in a contracted state. The number of quanta and the energy used in the continuous contraction of the sphincter can be measured or calculated based on the sphincter muscle's capacity to direct energy and on the information rate. Any given arthropod can have many tracheae and therefore many sphincters. The respiration of the arthropod depends on an information generator or generators sending information to the sphincters in a coordinated manner.

The respiration behavior of these simple animals also reflects the adjustment of the primary information rate to accommodate the sensing of carbon dioxide concentrations. Carbon dioxide sensors generate information which influences the primary information generator. However, it is the final output of the primary information generator that causes the breathing behavior. This primary information can be quantified and can be used to quantify the resulting breathing behavior.

As previously described, some arthropods have tracheae pumps in addition to the spiracles and sphincters described above. These are simple pumps consisting of muscle tissue of the tracheae and the motoneurons associated with this muscle tissue. The behavior of tracheae pumps can be quantified from the characteristics of the pump (i.e., the number of motor units, the type of muscle tissue, and the muscle configuration), from the information rate of the primary information generator, and from the amount of energy available.

7.4.5.2. Mammals. At the high end of the respiratory systems' complexity scale are the mammals. This complexity arises due to the large number of motor units associated with the respiratory pumps of these large animals and to the many types and numbers of secondary information generators which can influence these motor units. As previously described, both the diaphragm and the intercostal muscles are components of the respiratory pumps of mammals. Each of these components contains a number of motor units which provide them with a wide range of behaviors. For example, in shallow breathing, the chest cavity expands very little because only a few motor units in the diaphragm and chest muscles contract. During very deep breaths, the chest cavity expands a great deal because many of the motor units in the diaphragm and the chest muscles are contracting.

The respiratory system's behavior is at a minimum under the conditions of basal metabolism. Under these conditions, the body is at rest, the ambient environmental conditions (e.g., temperature), are standard, and sufficient time has passed since eating so that the digestive tract is in a low activity state. For basal metabolism, both the information flow rate and the work done by the breathing muscles are at the minimum level compatible with maintaining life. For any specific animal, the information flow rate and the work performed by the respiratory pump can be quantified.

Under conditions of active behavior, such as exercise, both the depth and the rate of respiration are increased from those of basal metabolism. This increase is caused by an increase in the rate of information from the primary respirator information generator and by the distribution of information to more respiratory muscle motor units. In turn, this increase in rate of information and number of motor units activated is caused by one or both of two information-generating mechanisms. We first consider nervous information and then chemical infomation.

Nervous information regulates respiration as illustrated by an experiment devised by Krogh and Lindhart. In this experiment, the subject is placed on a stationary bicycle ergometer with a special type of brake, the actual work done being measured by the brake. In this case the brake was an electric motor. The resistance to the worker and hence the work which was performed in overcoming it could be regulated by adjustment of the current passed through the motor. When work was commenced there was an immediate increase in the rate of depth of respiration and also in the pulse rate. Had these alterations been due to the stimulating action of chemical products formed in the muscle on the respiratory center, enough time must have elapsed to allow for the products being taken up by the blood, carried to the heart, passed through the lungs, and then to the brain. These processes would have occupied up to half a minute. In fact, the augmentation of respiration came about much more quickly—in about five seconds from the commencement of exercise.

In a similar experiment using the same stationary bicycle as before, the subject was led to believe that the load on the machine (and consequently the exercise he was to take) was suddenly increased by throwing in a powerful current. Actually the current was not increased, although the actions of closing the switches, etc., were performed. The pulse and respirations were augmented as before, although no extra work was done by the subject. Clearly, therefore, the increased respiratory efforts were not due to chemical products produced by the work. The neural information in this second experiment was generated in the higher levels of the central nervous system (the brain) since it involved seeing what appeared to be an increased, and anticipated, work demand. The first part of the experiment determined that other neural information generators can change the output of the primary respiratory information generation and its distribution.

The chemical regulation of breathing behavior was first clearly demonstrated by Haldane and Priestly (1922) based on their discovery of a method of determining the composition of alveolar air. They found that at various altitudes (air densities) the actual partial pressure of carbon dioxide in the alveoli, and hence the concentration of that gas in the blood, remains almost unchanged. This constant pressure of carbon dioxide in the alveolar air means that the greater the air density, the more the carbon dioxide produced by the body is diluted in the lung. As the carbon dioxide produced by the body is approximately constant in amount, the total ventilation of the lung, and hence the respiratory efforts, must be greater when the air density is higher, for example, at lower altitudes. Any effort to increase the carbon dioxide pressure in the blood, as by inhaling carbon dioxide or diffusing it into the blood as the result of muscular exercise, has the effect of stimulating the respiratory center and increasing the respiratory efforts, especially the depth of respiration. Clearly, chemical sensors generate information that changes the neural information that causes breathing behaviors.

7.4.6. Quantification of Respiratory Behavior

The basic concept of breathing behaviors is simple for those animals with respiratory systems consisting of pumps and neural systems. The breathing behavioral system consists of (1) mechanical pumps constructed of muscle, which in turn consists of motor units composed of a motoneuron and its associated muscle fibers; (2) a primary information generator and distribution system which provides information to the motor units; (3) secondary information generators which can modify the output of the primary information generator; (4) a chemical energy source which is available to the motor units for conversion to heat and work; and (5) the behavior which results due to the operation of the pumps. This basic concept provides the framework for quantification of the parameters associated with behavior and for the quantification of breathing behaviors.

The mechanical respiratory pumps of all animals exhibit the same fluid-pumping behavior. The amount of fluid they can pump, that is, their capacity to direct energy, is a function of their structure, as previously described. The amount of energy a pump can impart to a fluid per quantum of information can be quantified using Eq. 7.1 or 7.2. Of course, the amount of energy a specific pump can impart to a fluid is a function of the structure of that pump. For example, the amount of energy an arthropod can impart to a fluid is significantly less than that which can be imparted by a large animal such as a human. However, irrespective of size and structure, the energy capability of a respiratory pump can be quantified.

Respiratory pumps convert chemical energy into both work and heat. Therefore, it is necessary to determine the amount of heat generated for each contraction of the pump in order to determine the amount of chemical energy used in the contraction. This heat energy can be measured for each specific pump. The heat energy per contraction plus the work performed during each contraction of a pump is the amount of available chemical energy required per contraction. The amount of available chemical energy required for a given period, such as a day, can be obtained by multiplying the amount of chemical energy that must be available per contraction by the number of contractions in the specified period of time.

The information causing a particular respiratory behavior is a function of the primary information generator and the neural distribution system from this generator to the motor units of the respiratory pump. As described previously, a few quanta of information per inhalation and exhalation of a specific respiratory system may result in a shallow breath, whereas many quanta may result in a deep breath. For a particular respiratory system, it is possible to quantify the amount of information for a specific breath. A time history of breathing information can be obtained from the individual breath information.

Respiratory behavior can be quantified using the fundamental parameters as described in this chapter and the behavioral equations described in Chapter 5.

There are significant differences between the respiratory information-generating systems of simple and complex animals. These differences are due to the number and types of secondary information generators which can influence the primary respiratory information generator. Simple animals may have only a few sensors which are capable of generating information that can influence the primary information generator, whereas complex animals may have many more such sensors. However, it is the final output of the primary respiratory information generator that is the determinant of behavior.

The data and tests presented above are those necessary and sufficient to validate the concept of breathing behavior depicted in Figure 7.2. It can be concluded then (1) that the principles of a quantitative living systems science are applicable to the respiratory behaviors, and (2) there is the highest level of confidence in the application of these principles to respiratory behaviors.

7.5. The Digestive System

Most animals digest food extracellularly. The few that do not are multicellular animals (e.g., internal parasites such as tapeworms) with no digestive system. Such species live in an environment so rich in already digested nutrient molecules that they can just absorb them directly into their cells. Except for these few, most animals take food into a body cavity that is continuous with the outside environment. They then secrete digestive enzymes into that cavity, thereby reducing the food into nutrients which can be absorbed by the body. The behaviors of animals with digestive systems are considered here.

The simplest digestive system is a gastrovascular cavity that connects to the outside environment through a single opening. The guts of all animal groups other than sponges, cnidarians, and flatworms, are tubular, with an opening at either end. The data relating to the autonomous mechanical characteristics of animals with tubular guts were used to test the behavioral concept of digestion described early in this chapter.

7.5.1. Digestive Structures

Different regions in a tubular gut are specialized for particular functions. At the anterior end of the gut are the mouth (the opening itself) and buccal cavity (mouth cavity). Food may be broken up by teeth (in some vertebrates) or by mandibles (in insects), or somewhat further along the gut by structures such as the gizzards of birds and earthworms, where muscular contractions of the gut grind the food together with small stones. Some animals simply ingest large chunks of food with little or no fragmentation. Stomachs and crops are mainly storage chambers, enabling animals to ingest relatively large amounts of food and digest it at leisure. Digestion and absorption may or may not occur in the stomach and crop, depending on the species. The next section of the gut is the midgut or intestine. Food delivered into this section is well minced and mixed. Most digestion and absorption occurs here. Specialized glands secrete digestive enzymes into the intestine, and the gut wall secretes other digestive enzymes. The hindgut is the next section. It often includes a muscular rectum near the anus. This hindgut recovers water and retains feces.

The autonomous behaviors of the digestive system are those associated with the movement of food through the gut. These behaviors are caused by gut muscles and the autonomous nerve system. The structural plan of the gut involved in these autonomous behaviors is described in the concept of digestion presented earlier.

Although the specific structures of the guts of the various species are very different in appearance, they are similar in function. This wide range of specific structures is illustrated by the straight tube of the Nematode gut to the complex gut of vertebrates with their esophagus, stomach, small intestine, and large intestine. However, the function of the structure of the gut is to propel the contents of the gut from the start of the gut to its end. In effect, the gut is a linear pump which squeezes the semisolid contents. This pumping effect is achieved through the circular and longitudinal muscles of the gut. Contraction of the circular muscles reduces the internal cross-sectional area of the gut and contraction of the longitudinal muscles reduces the length of the gut. The combined contraction of the circular and the longitudinal muscles provides a "milking" action which squeezes the food products through the gut.

7.5.2. Capacity to Direct Energy

The structure of the gut pump is different from those of the heart and breathing pumps. The muscles of the gut are the smooth muscles, whereas the muscle of the heart is of a different texture. The skeletal muscles associated with breathing pumps are the striated muscles. Compared to the speed of contraction of heart and skeletal muscles, the smooth muscles of the gut contract slowly, resulting in a slow squeezing of the gut and a slow progression of the food materials in the gut. The structure of the gut can be considered as a series of elemental pumps along the entire length of the gut. The contraction of the muscles of one elemental pump squeezes the materials in that pump onto the next elemental pump.

The work performed by the gut muscles is a function of the mass of the material moved in the gut, the resistance the material is moved against, and the distance the material is moved during each muscle contraction. The work performed by any specific gut pump is a function of the characteristics of that gut.

As an illustration, the cross-sectional area, length, and number of compartments of the earthworm's gut are very much different from those of the human gut. To obtain the capacity to direct energy, or capacity to perform work, of a specific gut of a given species, the characteristics of the gut pump must be specified.

7.5.3. Available Energy

The smooth muscles of the gut, like the heart and skeletal muscles, must have chemical energy available to them for conversion to mechanical work and heat. The form of the chemical energy that must be available to the muscles of the gut is the same as that for the other types of muscle.

7.5.4. Neural Information Generation and Distribution

The gut, like the heart, can generate its own information, transmit this information to its muscle tissue, and cause the muscle tissue to contract—the gut has its own intrinsic nervous system. Neural information can travel from one region of the digestive tract to another without being processed by the central nervous system. For example, loading the stomach with food can lead to muscle contraction behavior associated with a bowel movement.

The primary information generators associated with the gut's autonomous behavior are the sensors for determining the stretched condition of the gut. These information generators operate in a coordinated way so that the activities in various regions of the gut act in a properly timed, assembly-line manner. It is this coordination of the information generators that is responsible for peristalsis, the wave-like phenomenon which squeezes the contents of the gut causing the material in the gut to slowly move through. The primary information generators are not directly connected in the sense that one generator does not send information directly to the next generator along the gut. Instead, an information generator causes the gut muscles associated with this generator to contract, which, in turn, causes the stretch sensor in the contracted muscle to generate information which is then transmitted to the next set of muscles in the gut. It is this successive generation of information–muscle contraction–information generation that causes peristalsis.

7.5.5. Digestive Behavior

The behavior of the digestive system includes all those activities necessary to convert raw food into a form that can be used by the cells of the body along with the elimination of the byproducts of this food processing. These activities include (1) the volitional muscle behaviors of shredding, pulverizing, filtering, chewing, or otherwise processing raw food, (2) the volitional muscular behavior of eliminating food, (3) the autonomous chemical information related behaviors, and (4) the autonomous neural muscle behaviors. Only the last item is considered in this treatment of autonomous behaviors caused by neural information.

The autonomous digestive behaviors are initiated with the swallowing of the food in the mouth. At this point, the digestive behavior is the movement of food through the gut while the food is being chemically and mechanically processed and the movement of the byproducts of this processing to the end of the gut. In essence, the autonomous digestive behavior caused by neural information is that of a series of muscular pumps. As with any mechanical pump, the work done by the pump can be measured or calculated when the characteristics of the pump and of the matter being pumped are known.

Although in the most general terms the gut is a single tube running through the body from the mouth to the anus, for specific species there may be compartments specialized for digestion and absorption. The gut of the Nematode is simple and consists of an intestine running from the mouth to the anus without any specialized compartments. The earthworm has a pharynx, an esophagus, a crop, a gizzard, and an intestine, each with different muscular structures. The snail has a stomach and an intestine. The cockroach has an esophagus, a crop, a gizzard, an intestine, and a rectum. The rabbit has an esophagus, a stomach, a large intestine, a cecum, a small intestine, and a rectum. The behavior of a particular gut is a function of the number of compartments and the characteristics of each compartment.

The muscle contraction behavior of any compartment is a function of the muscle configuration, available energy for the muscles, and the information and its distribution to the muscles. The behavior of the total gut can be determined from the behavior of each of the compartments in response to the information that causes the muscles of each compartment to contract.

7.5.6. Quantification of Digestive Behavior

The foregoing data are sufficient to demonstrate that the general concept for the quantification of digestive system behavior is valid and follows the same plan as the autonomous behaviors of the cardiovascular and the respiratory systems. That is, a given system's capacity to direct energy can be quantified based on its structures and organization, the amount of chemical energy that must be available for the behavior can be determined, and the information necessary to cause the behaviors can also be determined. The fundamental autonomous behavior of digestion can be quantified based on the equations developed in Chapter 5 and on the quantification of the fundamental parameters—capacity to direct energy, available energy, and behavioral information.

7.6. Summary and Conclusions

The most easily observed, constant, and consistent behavioral phenomena of animals provide the basis for verifying the principles of a quantitative living systems science as they apply to the behaviors of animals. These phenomena are the autonomous behaviors of the cardiovascular, respiratory, and digestive systems. Fundamental concepts for quantifying the behaviors of these systems were developed in terms of the fundamental determinants of behavior, that is, a capacity to direct energy, available energy, and behavioral information. Each of these fundamental determinants can be quantified, thus allowing quantification of the behaviors of these systems.

There is a robust literature describing the characteristics and the autonomous behaviors of these three systems. The applicable data from this literature were used to test the fundamental concepts for quantifying the basic behaviors of the cardiovascular, respiratory, and digestive systems. These data verify the validity of the fundamental quantitative behavioral relationships postulated in the behavioral concepts.

Testing the principles of quantitative living systems science using data on the autonomous behaviors of animals results in the following conclusions:

1. The autonomous behaviors of animals are easily described in terms of the principles of quantitative living systems science.
2. The autonomous behaviors of animals can be quantified using the principles of quantitative living systems science.
3. The demonstrated ability of the principles of quantitative living systems science to quantify the autonomous behaviors of animals results in a very high confidence level in the validity of these principles.

CHAPTER 8

Nonvolitional Behaviors

8.1. Introduction

In addition to the autonomous behaviors caused by internally generated information (treated in Chapter 7), there are autonomous behaviors caused by information generated as a function of environmental phenomena. These externally caused behaviors are called *nonvolitional* to differentiate them from the autonomous behaviors caused by internally generated information. These nonvolitional behaviors include those known as stimuli–response behaviors. Testing and validation of the principles of a quantitative living systems science as they apply to nonvolitional behaviors are the subjects of this chapter.

The general models, processes, and procedures used in earlier chapters are again used here for the testing and validation of the principles of a quantitative living systems science. Thermal phenomena in the environment are treated first because they are (1) always present, (2) readily observable, (3) relatively consistent over large areas of the earth, and (4) the cause of readily observable nonvolitional behaviors of animals. For example, when we are hot we automatically perspire, and when we are cold, we shiver and cannot willfully stop the shivering. Gravitational phenomena are treated next because they are ubiquitous, they are readily observable, they are consistent, and some behavioral responses to these phenomena are nonvolitional. The next phenomena considered was light, which, although neither constant nor consistent, affects most animals by causing nonvolitional behaviors. The last phenomena considered are the nongravitational mechanical forces in the environment which can cause nonvolitional behaviors such as the "knee-jerk" reaction.

Animals' sensors provide them with a capability to generate behavioral information about some or all of these environmental phenomena. This information causes nonvolitional behavioral responses to these environmental phenomena.

The concepts developed in preceding chapters, along with data on nonvolitional behaviors, were used to synthesize general concepts of behavior and information as they relate to the nonvolitional behaviors of animals. These concepts were then tested against validated data in order to establish the level of confidence of the concepts. After validating the general concepts for the nonvolitional behaviors, these concepts were further defined and detailed.

8.2. Concepts

The concepts of the autonomous behaviors given in Chapter 7 can be extended to include those nonvolitional autonomous behaviors which are caused by phenomena external to an animal. This is achieved by considering the ways the motoneurons and motor units are related to the phenomena in an animal's environment. In general, the way an animal's neural system is connected to the animal's environment is through its sensors. In the special case of nonvolitional behaviors, nerves go directly, or essentially directly, from the sensors to motoneurons of the behaving motor units. From a quantitative living systems science perspective of nonvolitional behaviors, sensors interact with phenomena in the environment and generate neural information which is distributed directly to motoneurons and their motor units causing the nonvolitional behaviors. The concepts for nonvolitional behaviors resulting from environmental phenomena are presented below. These phenomena are sufficient to demonstrate the concept of autonomous behaviors based on environmental phenomena, or nonvolitional behaviors.

8.2.1. Thermal Phenomena

All animals respond to the thermal environment in some way. For example, below some level of thermal energy in their environment animals die. For most animals, a body temperature below the freezing point of water means death. In the case of warm-blooded animals, such as the human species, a body temperature only a few degrees lower or higher than the "normal" results in death. Within the "normal" range of temperatures, animals have autonomous behaviors which are related to their thermal environments. As mentioned above, the easiest to observe of autonomous thermal behaviors is perspiring due to elevated temperatures and shivering due to cold temperatures. The concept of this behavior is that changes in the thermal environment cause sensors at the surface of the body to generate neural information that is distributed to muscles in the skin. In the case of perspiration, this information causes skin pores to open, and in the case of shivering it causes muscles to contract and relax, which, in turn, generates heat in muscles.

The concept of this nonvolitional behavior in response to the thermal environment is shown pictorially in Figure 8.1. This concept has been tested using nonvolitional behavior data and is described later in this chapter.

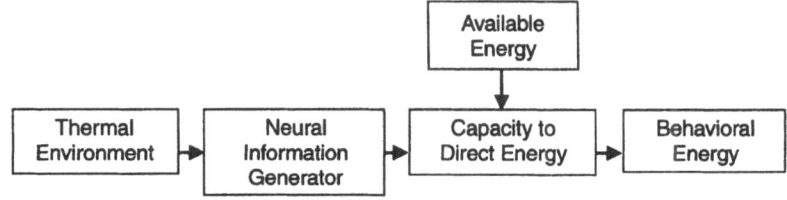

Figure 8.1. Behavioral response to the thermal environment.

8.2.2. Gravity Phenomena

The gravity environment in which we live is essentially constant, making behavioral responses to gravity phenomena difficult to observe. However, animals do have autonomous behaviors that are related to gravity phenomena. Animals have an upside and a downside—a characteristic related to gravity that was not understood for a long time. We have gravity sensors that are capable of differentiating small changes in our body's relationship with respect to gravity. For example, the inner ears are sensitive to small differences in the effects of gravity on the body. Also, kinesthetic senses can determine muscle tension with respect to gravitational effects. These sensors generate neural information which is distributed to appropriate motoneurons and motor units to cause them to behave in ways that align the body with respect to gravitational forces.

The concept of nonvolitional behaviors in response to gravity phenomena is shown pictorially in Figure 8.2. This concept has also been tested using nonvolitional behavioral data and is described later in this chapter.

8.2.3. Light Phenomena

The light environment results in nonvolitional behaviors in humans and in some other animals. A readily observed nonvolitional behavior in response to the light environment is the changes in the pupil of the eye caused by increased or decreased levels of light energy. This change in pupil size is of interest because

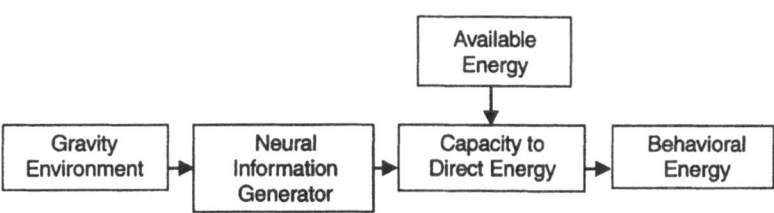

Figure 8.2. Behavioral response to gravity phenomena.

the sensing element, distribution, and muscle response are localized in the eye and do not involve differentiated nerves. The concept for behaviors associated with light phenomena is depicted in Figure 8.3.

8.2.4. Mechanical Environment

We live in an environment that can impart mechanical forces on our bodies. These forces range from the gentle breezes that we enjoy to the deadly blows by predators and enemies that can kill us. These forces also include the mechanical blows that stretch tissue in our legs and cause the well-known "knee-jerk" behavioral reaction. This concept is depicted in Figure 8.4. This concept was also tested using nonvolitional behavioral data to determine the concept's validity.

8.3. The Thermal System

The metabolic functions of animal cells are mostly limited to the range of temperatures between 0°C and 45°C, which can be taken as the thermal limit of life. Therefore, they must have ways to maintain their cells within the thermal limit of life. Animals have various methods for achieving temperatures within this range. These methods include those autonomous behaviors associated with neural information and muscle behavior. In keeping with the approach of first treating those behaviors which are most obvious and easily observed, shivering behavior is considered first. Humans exposed to low temperatures experience shivering behavior in response to these low temperatures. Shivering behavior is widespread among mammals and birds. The less obvious autonomous behavior of varying the amounts of blood exposed to the environment via the skin is also considered. Data related to these behaviors were used to test the thermal concept described earlier in this chapter. The processes used to test the cardiovascular, respiratory, and digestive systems are also used to test the thermal system.

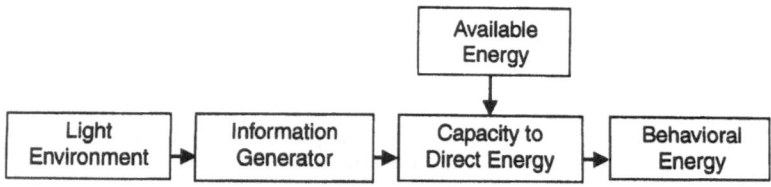

Figure 8.3. Behavioral response to a light environment.

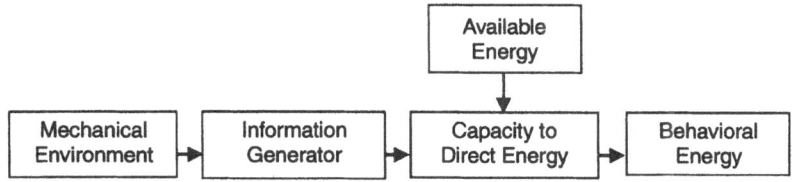

Figure 8.4. Behavioral response to the mechanical environment.

8.3.1. Thermal System Structures

Organisms have numerous thermal systems for maintaining optimal body temperatures. The common classifications are cold-blooded and warm-blooded animals. However, biologists prefer three different classifications which are more descriptive: (1) homeotherm, an animal that regulates its body temperature at a rather constant level; (2) poikilotherm, an animal whose temperature changes; and (3) heterotherm, an animal that regulates its body temperature at a constant level some of the time. Another thermal system classification scheme is based on the type of thermoregulatory mechanisms. Ectotherms are animals that depend largely on external sources of heat, such as solar energy, to maintain their body temperatures above the environmental temperature. Endotherms are animals that can produce substantial heat energy through the metabolic process and can maintain body temperature by regulating heat loss to the environment. Mammals and birds behave as endotherms whereas animals of other species behave as ectotherms most of the time.

Endotherms were selected for testing the thermal system concept because our primary interest is in humans and because many of the behaviors of endotherms are autonomous. The basic structure of the autonomous thermal system of endotherms consists of (1) a primary thermal information generator located in the hypothalamus of the brain; (2) an information distribution system that distributes information to the behaving elements; (3) behaving elements consisting of (a) the smooth muscles of the blood vessels that can constrict these vessels and restrict blood flow and (b) paired skeletal muscles; and (4) the skin temperature sensors, which are information generators responsive to environmental temperatures.

8.3.1.1. Skeletal Muscle Structures. Shivering behavior is caused by the skeletal muscles. The structure of skeletal muscles that allows shivering is the paired nature of these muscles. Movement of one body part with respect to another part is the result of two muscles (a muscle pair) operating in a coordinated way to rotate a body part around a joint between two body parts. To obtain movement, one muscle of the pair contracts and the other relaxes. Shivering behavior results when both of the muscles of the pair contract—there is no resulting mechanical motion and the energy of muscle contraction is converted to heat energy. The heat energy resulting from shivering is used to heat the animal's body.

8.3.1.2. Smooth Muscle Structures. Heat exchange between an endotherm's body and the environment is a function of the amount of blood that is flowing in the animal's skin, which is a function of the diameter of the vessels supplying blood to the skin. The diameter of these vessels is determined by contractions of the smooth muscles which are integral to the blood vessels. The more the contractions, the smaller the diameter of the blood vessels, the smaller the amount of blood in the vessels, and the less heat flowing from the animal to the environment. In a cold environment, the blood vessels are constricted while in a hot environment, the vessels are dilated.

8.3.2. Capacity to Direct Energy

Two characteristics of skeletal muscle—the organization into opposing pairs of muscle and the ability to contract these muscles simultaneously—provide a capacity to direct energy in the form of heat rather than the normal alternate contraction of the muscles which produces both work and heat. When contracting isometrically, the paired muscles pull against each other producing no skeletal movement and only slight muscle movement in the form of tremors. The amount of heat energy generated during isometric contractions can be measured.

By virtue of the smooth muscles in the blood vessels, animals have a capacity to direct energy which results in the contraction of these muscles. Contraction of these blood vessel muscles change the amount of blood flowing in the vessels and thereby indirectly directs the amount of heat energy that is lost to the environment. As with all muscle tissue, the amount of total energy used in the contraction of this tissue can be either measured or calculated.

8.3.3. Available Energy

In both skeletal muscle and smooth muscle contractions for an animal dealing with its internal and external thermal environments, chemical energy in the form of ATP is converted into ADP. As described in previous chapters, there must be sufficient chemical energy available to these muscles for them to convert to heat energy and work.

8.3.4. Neural Information Generation and Distribution

The primary behavioral information generator for thermal behavior of vertebrate animals is in the brain's hypothalamus. The hypothalamus is located at the bottom of the brain where it is associated with the distribution of the motor nerves to the skeletal and the blood vessel muscles. The hypothalamus determines the "core" temperature of a vertebrate, sometimes referred to as the set point for the metabolic heat production response. The hypothalamus is to thermal behavior as the pacemaker is to cardiovascular behavior and the pons in the brain stem is to respiratory behavior.

Under conditions of basal metabolism, the temperature of the hypothalamus has a particular temperature value. If the temperature of the hypothalamus of a mammal is decreased below this value, the vessels supplying blood to the skin constrict. A low hypothalamic temperature will initiate shivering. A hypothalamic temperature two or three degrees higher than normal will initiate panting and sweating in some animals.

Under conditions other than the basal metabolism temperature, the skin temperature sensors generate neural information which is transmitted to the hypothalamus where it changes the information output of the hypothalamus. Changes in skin temperature shift the hypothalamic set points. The set point for the metabolic heat production response is higher when the skin is cold and lower when the skin is warm. It is the information from the skin sensor generators that, when transmitted to the hypothalamus, adjusts the hypothalamic set point. Other nonvolitional factors also shift hypothalamic set points for responses. Set points are higher during wakefulness than during sleep. They are higher during the active part of the daily cycle than during the inactive part, even if the animal is awake at both times. Details on these phenomena are provided by Purves *et al.* (1992).

8.3.5. Thermal Behavior

The wide range of thermal environments over the earth and the narrow range in which animals can exist have resulted in many adaptations for coping with these conditions. However, the nonvolitional behaviors are limited to changes in the metabolic rate and the heat transfer rate of animals. The metabolic rate is related to the amount of chemical energy that is converted into heat energy and the heat transfer rate is related to the loss of heat to the environment by changing the heat flow rate from the animal to the environment. These two schemes have evolved into the three major classes: ectotherms, endotherms, and heterotherms. These three classes have some variations, such as hibernation and shallow torpor.

8.3.6. Quantification of Thermal Behavior

It is theoretically possible to determine the structure of the nervous system from the primary thermal information generator to the motor units associated with the nerves of this system. And it is also theoretically possible to determine the amount of heat energy generated in the motor units associated with shivering.

With the nervous system and the motor units identified and characterized, it is possible to calculate the amount of heat energy generated for each quantum of thermal information transmitted to these motor units. In other words, the structure of the thermal information generator, the information distribution, and the behavior elements (motor units) can be identified and characterized, and the behavior of these systems can be quantified. An alternative method for quantifying shivering behavior is to measure the total increase in heat generated from shivering and to measure the number of quanta that cause the muscle contractions which generate this increased heat.

It is also theoretically possible to determine the structure and characteristics of the information generator, the information distribution, and the smooth muscles of the blood vessels. From this determination, the behavioral energy in the contraction of the blood vessels and the information which causes these contractions can be quantified. It is also possible to determine the change in heat flow from the blood to the environment and the information associated with this change in heat flow and thereby quantify heat transfer behavior.

Although the specific autonomous thermal behaviors of animal species vary greatly, the general scheme for these behaviors can be characterized by the conceptual system depicted in Figure 8.1. The data verify the validity of this conceptual system.

8.4. The Gravitational System

All things on earth are in a gravity environment. Both plants and animals have biological mechanisms for sensing gravity phenomena. In plants, the behavior related to gravity is called gravitropism, the growth to or away from the gravity stimulus. In animals, there are varied types of behavioral responses to gravity phenomena that are associated with equilibrium and balance. Almost all animals, whether at rest or in motion, tend to assume some definite bodily position, which is called the normal position, with respect to the force of gravity and other forces acting upon them. Normal posture is achieved by constant muscular activity. This activity is controlled by the central nervous system which receives information from the various sense organs and combines this information to cause the necessary position adjustment. There are several kinds of information, including cues from the eyes, contacts with surfaces and objects, and mechanical displacements of internal organs or appendages, but many animals receive information from a sensory apparatus developed especially for the sense of equilibrium. The structures which serve this sense differ in form and complexity in invertebrate and vertebrate animals.

In general, the mechanisms for sensing gravity in animals are mechanosensor hair cells and the stretch sensors in muscle spindles which indirectly sense gravity via unequal stretching of muscles. These sensors of gravity generate the primary information used for equilibrium behaviors. The data on gravitational sensors and muscular structures were used to test the validity of the basic autonomous gravity behavior concept presented earlier in this chapter.

8.4.1. Gravity System Structure

The primary structure for animal equilibrium is the hair sensors that generate information in response to gravity/body relationships, the information distribution system, the skeletal structure, and the skeletal muscles. Since there are significant differences in the structures for sensing gravity phenomena and for establishing equilibrium invertebrates and vertebrates, their structures are described separately below.

8.4.1.1. Invertebrates. The primary gravity sensor/information generator in many lower animals is in an organ called a *statocyst*. The statocyst is approximately spherical in shape, is lined with sensory cells, and contains an internal movable mass, the *statolith*. It is this mass which moves in the organ in response to gravity forces. When the body is in the normal position the statolith rests against the lowermost part of the organ, in contact with a particular group of sensory cells. When the body position is changed, the statolith is displaced to a new position where it activates other hair sensors which generate information that is transmitted via the nerve fibers to the muscles of the legs, wings, or fins, whose action then tends to restore the animal to the normal position. The specific structure of the statocyst varies somewhat among animals due, in part, to the density of sensor hairs in the organ. Those with high density provide a high degree of precision in adjustment to gravity forces. For example, experiments on the slug show that this animal can detect a tilt of its body of as little as one-third of a degree. Ctenophores, mollusks (such as bivalves, snails, and squids), and crustaceans all have statocysts of the type described above. Some animals, like the jellyfish *Aurelia*, have statocysts of a different structure that are called tentaculocysts. They are eight club-like projections, weighted at the end with crystals, arranged around the margin of the umbrella; they assist in the maintenance of an upright position during swimming. The structure of the gravity sensors is such that deviation of the body position with respect to the vertical gravity vector generates information about the direction and magnitude of this deviation.

The structure of the distribution of the information generated in the sensor is such that the information generated in the sensitized part of the sensor is distributed to those muscle fibers which will exert appropriate mechanical forces that will restore equilibrium. The type of force depends on the skeletal and muscle structure of the particular organism. For example, animals without rigid skeletons, such as the jellyfish and the squid, use differential muscle contractions which change the velocity vector of their water jet propulsion to provide their capability for equilibrium restoration. For invertebrates, these structures are less complex than those of the vertebrates.

8.4.1.2. *Vertebrates.* The primary equilibrium organ of vertebrates consists of a number of sensory endings within the *membranous labyrinth*, a complex enclosure usually surrounded by bone or cartilage. In general, there are two sacs, the *utricle* and *saccule*, and three *semicircular canals*, all filled with a fluid. The utricle and saccule contain one type of ending, the *macula*, which somewhat resembles the statocyst of invertebrates. The semicircular canals have a different structure, the *ampullary crista*.

The macular endings form a patch on the inner surface of the membranous cavity that is approximately oval in shape. It is made up of many epithelial cells provided with short cilia which extend toward the interior. Over the cells and firmly embedding them is a dense gelatinous blanket, the *statolithic membrane*, which is weighted with numerous calcareous particles, the *statoliths*.

Stimulation of the macula is similar to that of the invertebrate statocyst. However, its statolithic mass is located in one place in the organ. As the organ assumes different positions, the mass hangs differently, exerting varying tensions on the cells at its base and stimulating them. In effect, its action is similar to that within the statocyst.

The second type of ending, the crista, is found in an enlargement (*ampulla*) at one end of each semicircular canal. It differs from the macula in several ways. The epithelial cells lie on the surface of a ridge and have long hairs extending into a gelatinous mound called a *cupula*. There are no statolithic granules. The cupula extends all the way across the ampullary cavity to the opposite wall, but it is not attached to this wall.

The labyrinth, as described above, is typical in all vertebrates except the most primitive, the lampreys and hagfishes. In the lamprey the labyrinth has only two semicircular canals, and the hagfish only has one.

In addition to the labyrinth, the sense organs in the muscles, tendons, and joints and those responsive to pressure and touch are continually generating information related to gravitational phenomena. The former group generates information related to the position of the head and limbs in relation to the body and the amount of tension each muscle is exerting, which is indirectly related to the forces of gravity. The latter generates information regarding where weight is resting and, thereby, also is indirectly related to the forces of gravity.

Unlike the cardiovascular structure with one primary information generator and one responding muscle, or the respiratory structure with a single primary information generator with a few responding muscles, the gravity system structure has a number of information generators transmitting information to a large number of motor units. To illustrate, the primary gravity sensing organs, that is, the statocyst and the labyrinth, contain many sensing cells which generate information. It takes a number of sensors to determine the body's position with respect to the maximum gravity gradient. In addition, a number of motor units must be available to obtain a fine gradation in positioning the body within a few tenths of degrees to the maximum gravity gradient reference.

The skeleton and muscle structures of vertebrates are more complex than those of the invertebrates; this, in turn, increases the equilibrium complexity of vertebrates. This increase in complexity is due to the paired muscle and skeletal joint structures of vertebrates. The motion of one skeletal element with respect to another connected by a joint depends on paired muscles consisting of a flexor and an extensor muscle. This motion depends on one of the paired muscles contracting while the other muscle relaxes. A steady-state condition of equilibrium, such as the standing position of a human, depends on a continuous flow of information to both the flexor and extensor muscles associated with the standing position. This steady flow of information causes both of the paired muscles to be in a continuous state of contraction, which provides sufficient muscle force to keep the skeletal joints in a fixed position. Under these conditions of equilibrium, any motion requires an increase in the amount of contraction in one muscle and/or a decrease in the amount of contraction in the paired muscle. In turn, the flow of information to one muscle increases and/or the flow of information to the paired muscle must be decreased.

8.4.2. Capacity to Direct Energy

The autonomous equilibrium behaviors of animals in response to gravitational phenomena are the result of skeletal muscle contractions. These muscles provide animals with their capacity to direct energy associated with equilibrium. The skeletal muscles involved in equilibrium are a direct function of a particular species' structure and organization. For example, the major skeletal muscles associated with human species equilibrium are those of the neck, torso, legs, and feet.

The normal equilibrium position of animals requires continuous skeletal muscle contractions; however, these contractions represent only a fraction of these muscles' capacity to direct energy. For example, a human's leg muscle energy used to maintain a normal standing position is small compared to the energy used in these muscles when they are used to lift heavy weights, because fewer motor units are used for the former behavior.

It is theoretically possible to identify the motor units associated with the maintenance of the normal position, which is the nonvolitional behavior related to gravity, and to determine the capacity to direct energy of these motor units. Theoretically possible means that the capacity to direct energy can be calculated from principles and data, as was done in determining the capacity to direct energy of a motor unit of the sartorius muscle of Rana pipiens in Chapter 4. There, it was necessary to derive the motor unit's capacity to direct energy because adequate measurement data could not be found.

Direct measurement of the amount of energy directed by a single motor unit is difficult with existing measurement instruments due to the small size of a motor unit and to the very low levels of energy involved (microcalories of energy). Direct measurement could be made if single motor units could be isolated and instruments were developed for measuring the energy directed by a motor unit. Until such time, the capacity to direct energy of motor units must be derived by other methods, such as those described in Chapter 4.

The capacity to direct energy for some particular equilibrium behavior, such as standing, can theoretically be determined by identifying the motor units used in the behavior and calculating the sum of the amount of energy utilized by all the motor units involved. Another way is to measure or calculate the metabolism energy for the standing animal and subtract the basal metabolism energy—the difference is the amount of energy used in the standing behavior.

8.4.3. Available Energy

The amount of chemical energy that must be available for equilibrium behaviors can be determined for each muscle contraction and for specified periods of time, such as a day. Either of the two methods listed above for determining the capacity to direct energy can be used to calculate the amount of energy that must be available for specific behaviors.

8.4.4. Neural Information Generation and Distribution

The primary information generators related to gravity phenomena are the structures identified above as the statocyst of invertebrate animals and the vestibular apparatus of vertebrates. The physical phenomenon upon which these two information generators are based is the well-known principle of gravitational attraction between two masses. These two masses are, for the invertebrates, the earth and the statolith inside the statocyst. For vertebrates, it is the earth and the otoliths (calcareous mass concentration) in the vestibular apparatus. Due to the great mass of the earth, there is a large gravitational force on the statolith and otoliths that pulls them toward the earth along a path which is perpendicular to the earth's surface. In the statocyst chamber, the statolith is free to move and is drawn to the lowest (closest to the earth) point of the statocyst sphere. In the vestibular apparatus, the otoliths, which are constrained in a gelatin, distort the gelatin in which they are embedded due to the pull of gravity on them.

The invertebrate's statocyst is a chamber lined with hair cells (mechanosensors). When the hair cells are bent by the statolith, neurotransmitters are released to the sensory neurons associated with these hair cells, thus generating information. In vertebrates, the vestibular apparatus contains two chambers whose functions are like that of the statocysts of invertebrates. Hair cells line the floor of the chambers. The sensitive part of the hair cells, the stereocilia, is embedded in a layer of gelatinous material. On top of this layer are many otoliths which are granules of calcium carbonate. When the organ moves, gravity pulls on the dense otoliths which bend the stereocilia of the hair cell causing neurotransmitters to be released to sensory neurons, thereby generating information.

The information generated by the primary sensors of gravity and equilibrium is transmitted via the nerves from these sensors to the brain stem. Information from other equilibrium sensors may be combined with the primary gravity sensing information in the brain stem. However, in the absence of information from these other equilibrium sensors, the primary gravity sensor information still is transmitted to the muscles associated with equilibrium to produce something like a coordinated posture. Distribution of primary gravity sensing information to the muscles associated with equilibrium via the brain stem was graphically demonstrated in experiments conducted over a half-century ago. Removal of both the cerebrum and the cerebellum in animals as high in complexity as cats and dogs did not interfere with their balancing power. But no trace of any posture, normal or abnormal, remained if the brain stem was destroyed down to the level of the medulla. Destruction of the upper part of the brain stem resulted in greatly reduced posture capability, but something like a coordinated posture still existed.

Equilibrium information also is generated by secondary gravity sensors such as the mechanosensors. These sensors are specialized cells that are sensitive to mechanical forces that distort their membranes. The mechanosensors in muscles, tendons, and joints are stretch sensors. They generate information about the position of the parts of the body in space and the forces acting on them. The unequal gravity forces acting on muscles, tendons, and joints when the body is not in a normal position are sensed by the stretch sensors. These sensors generate information which is automatically transmitted to the brain stem where it is combined with the information from the primary gravity sensors. This combined information is then transmitted to the motor neurons associated with equilibrium behavior.

8.4.5. Nonvolitional Behavior Related to Gravity*

Nearly all animals, whether at rest or in motion, tend to assume some definite body position with respect to the force of gravity. Animals with normal gravity sensors are observed to have effective locomotion behaviors, such as running, creeping, flying, and swimming. However, animals with defective gravity sensors have defective locomotion behaviors. Y. Delage was the first to experimentally verify the relationship between the gravity sensors and equilibrium and locomotion. He destroyed the statocyst organs of an animal with a needle and afterward observed serious disturbances of orientation and movement; for example, the animal lay on its side or on its back, turned somersaults, ran in circles, and the like.

Another experiment further demonstrated the relationship between gravity phenomena and equilibrium behaviors. V. Hensen observed that prawns lose their statoliths when they moult and that they pick up sand grains with their pincers and place them in their statocyst. A. Kreidl took advantage of this practice in a clever experiment. He placed newly molted animals in filtered seawater which contained fine iron particles. The animals placed these particles in their statocyst in place of the normal sand grains. When a strong electromagnet was held over the animal in a normal position, the "iron statoliths" were attracted upward, and the organ was stimulated just as if the animal had been turned on its back. Accordingly, it turned over and remained upside down as long as the magnet was kept in position. In effect, the electromagnet and the iron particles provided a means to simulate changing the direction of the earth's gravitational field.

The nonvolitional equilibrium behaviors caused by stretch sensors located in the muscles, tendons, and joints have been understood to some extent since the early work of Sherrington and Magnus. Recall from previous considerations that muscles stretch as a function of varying loads caused by body weight and position with respect to the earth's gravity field. Therefore, a relationship exists between gravity and the muscle stretch sensor. This early work by Sherrington and later by Magnus demonstrated that information from stretch sensors is primary for posture, as is the information from the labyrinths. This work was performed by selectively disabling various parts of the nervous system, as described below.

If both the cerebrum and the cerebellum are removed and only the upper part of the brain stem is destroyed, the nervous system is still able to produce something like a coordinated posture. The limbs are rigidly extended and the trunk arched backwards so that the animal is in something like its normal standing position if it is placed on its feet. This condition s known as "decerebrate rigidity." In decerebrate rigidity the only postural behaviors which remain are of no use to the

---

* The pre 1950 experiments described in this section are based on the article "Equilibrium" and "Equilibrium, Sense of" in Encyclopaedia Britannica, Vol. 8, 1951.

animal. The extended position of the limbs is retained for hours, but the animal cannot stand by itself and makes no attempt to change its position if placed on its back or side. There is no doubt that the posture of decerebrate rigidity is brought about by the same nervous mechanism that causes the behavior of standing.

The stretch sensors in the neck muscles have a special influence on normal posture. If the head is moved from its normal position relative to the trunk, one or another set of neck muscles will be stretched, the stretch sensors in the muscle will generate information that is transmitted to the brain stem. The effect of this information can be seen clearly after the destruction of the labyrinths since these also generate information as a result of head movement. When the labyrinths are destroyed, bending the head into any position will modify the posture of the body and limbs, usually in such a way as to bring the trunk into line with the head. For example, if the head is bent upward, muscle rigidity increases in the forelimbs and decreases in the hind limbs so that the animal squats on its haunches with the body inclined upward in line with the head. If the head is rotated, the extension of the limbs increases on the side toward which the jaw is turned so that the body tends to rotate on its long axis in the same direction as the head.

The experiments just described demonstrate that the nonvolitional behaviors of animals in relation to gravity phenomena are a primary function of their gravity sensors and of the stretch sensors in their muscles. In turn, these behaviors are a direct function of the information generated by the gravity sensors and the stretch sensors. For just the minimum coordinated motion behavior, both these types of information are necessary.

8.4.6. Quantification of Nonvolitional Gravity Behavior

The nonvolitional behaviors of animals related to the earth's gravity can be quantified. First, it is necessary to determine an animal's capacity to direct energy related to gravitational forces. This requires the specification of the animal's structure and organization characteristics as they apply to (1) the muscles used in the behaviors related to gravity forces, (2) the skeletal structure to which these muscles are connected, (3) the sensors of gravity and the information generated by these sensors, and (4) the information distribution system. Given these characteristics, the animal's capacity to direct energy can be determined, the information generated identified and calculated, and the behavioral energy calculated from the equations presented earlier in this book.

Since there are two sources of information based on gravity forces and since this information is combined to cause the resulting equilibrium behavior, great care must be taken in how this information is combined. It is this combined information that is transmitted to the motor units that causes the behaviors.

Quantification of nonvolitional behaviors using the methods described above is sufficient evidence to validate the concept depicted in Figure 8.2.

8.5. Light Energy Responses

The most obvious nonvolitional behavior in response to light energy is the change in the pupils of eyes as a function of light energy intensity. Contraction and dilation of the pupils are automatic responses to increases and decreases of light energy intensity. We cannot change this behavior of our eyes. As a matter of fact, the light energy sensing, information generation, transmission of information, and contractile tissue are all contained in the muscle of the eye's iris. Therefore, the central nervous system is not involved in any way in this iris contraction behavior.

The contraction behavior of the pupil of the eye is of interest because the central nervous system is not involved in the behavior, thus providing more evidence of the direct relationship between a system's capacity to direct energy, information, and behavior.

8.5.1. Structure of the Eye

Human eyes, those of other vertebrates, and those of cephalopod mollusks have exceptional abilities to form images of the visual world. The vertebrate eye is a spherical, fluid-filled structure bounded by a tough connective tissue layer called the sclera. At the front of the eyeball, the sclera forms the transparent cornea through which light enters the eye. Just inside the cornea is the iris with its central aperture, the pupil. The iris controls the amount of light reaching the photosensitive tissue at the back of the eyeball. The iris is under the control of the autonomic nervous system.

The contraction of the pupil of the eye in bright light is a well-known fact and has long been recognized as an example of simple reflex behavior in which the input information is through the optic nerve and then through the oculomotor nerve to the contractile tissue of the iris (Parker, 1918). Nevertheless, the sphincter of the pupil also acts under direct stimulation. The experiments of Steinach (1892) in the last decade of the nineteenth century on the contraction of the iris of the frog and the eel when illuminated provided evidence of this direct stimulation behavior. Steinach concluded that in fishes and amphibians the smooth muscle elements of the sphincter pupillae may be directly stimulated by light.

Steinach's investigations were confirmed by the work of Hertel (1906), whose studies included not only the eyes of the lower vertebrates but also those of the higher forms, including man. Hertel illuminated the eye with various types of light after the optic nerve had been cut. The iris of the eye blinded by cutting the optic nerve contracted when stimulated by these light sources. Based on Hertel's work, the sphincter pupillae of the vertebrate eye, and probably also that of the cephalopod eye, may be regarded as muscles normally subject to direct stimulation by light, notwithstanding the fact that they are also under nervous control. As experimental methods and measurement instruments advanced, it became possible to determine the time between a light stimulation and the response of the iris. The actual iris response time and the time necessary for an iris response if the stimulus was transmitted through the optic and the oculomotor nerves are different, resulting in the conclusion that the iris can be stimulated directly by light energy.

From the experiments just cited, it can be concluded that there are two structures associated with the nonvolitional behavior of iris contraction. One is the light energy sensor which generates information that is transmitted to the central nervous system via the optic nerve, then from the central nervous system via the oculomotor to the sphincter muscle of the iris. The second is the light energy sensor that generates information that is transmitted via muscle tissue directly to the muscle fibers in the iris sphincter.

8.5.2. Capacity to Direct Energy

There is a particular capacity to direct energy for each of the structures identified above for the nonvolitional behaviors of the iris in response to light energy. These capacities to direct energy are also specific for each species. The capacity to direct energy for a given species is determined by the number of muscle fibers in the iris, the way these fibers are structured to form the iris, and the number of muscle fibers that can be activated by the direct information input and by the information that is transmitted to the fibers via the optical and oculomotor nerves.

To determine the capacity to direct energy associated with the iris, it is necessary to identify the elemental behaving structure, just as it was necessary to identify the elemental motor unit behaving structure of the behaviors previously treated. Identification of the elemental behaving structure (i.e., light energy sensor, optic nerve, oculomotor nerve, and iris muscle fiber) is done with the same procedures used to identify the skeletal motor unit. That is, the elemental behaving structure is a single motoneuron (the oculomotor neuron) and the iris muscle fibers that can be activated by this single neuron. The presence of a number of motor units in the iris is evidenced by the fact that gradations in the size of the pupil, which is controlled by the iris, are almost imperceptibly small. The all-or-none characteristic of motor units means that a large number of motor units are required to obtain smooth changes in the pupil's size.

Just as with other muscles, the energy used in the contraction of the iris' muscles can be determined either by direct measurement or by calculation. Because there is an information input which causes these muscles to contract, the capacity to direct energy of iris muscles can be calculated based on an information input.

The structure for the direct stimulation elemental behaving unit is not as clear as that for the more typical oculomotor neuron/muscle fiber element just described. In some instances, it may be possible for the light energy sensor to be incorporated in the muscle fiber it controls. The measurement of energy levels associated with individual muscle fibers may be very difficult with today's technology. However, it is theoretically possible to identify the elemental behaving unit at this level of structure and to determine the capacity of this structure to direct energy.

8.5.3. Available Energy

The energy available for the behaviors of the iris contractions is the chemical energy that is converted by the iris' sphincter muscle. In essence, it is the same type of chemical energy that is available for the behaviors of the skeletal and the smooth muscles of the body.

8.5.4. Neural Information Generation and Distribution

Photosensitivity (sensitivity to light) is a characteristic of most animals. The basis for photosensitivity is the same across the whole range of animal species—the molecule rhodopsin. Photosensitivity depends on the ability of a rhodopsin molecule to absorb photons of light energy and to change its conformation. Rhodopsin consists of a protein, opsin, which by itself does not absorb light, and a light-absorbing group, 11-*cis* retinal. This light-absorbing group is located in the center of the opsin, and the entire rhodopsin molecule sits in the plasma membrane of a photosensor cell. When the 11-*cis* retinal absorbs a photon of light energy, its shape changes so that it becomes a different isomer of retinal, all-*trans* retinal. This conformational change puts a strain on the bonds between retinal and opsin, and the two components break apart. This disassociation of retinal and opsin causes the molecule to lose its photosensitivity. The retinal spontaneously returns to its 11-*cis* isomer and recombines with opsin to become, once again, photosensitive rhodopsin.

The rhodopsin molecule sits in the membrane of a photosensitive cell, such as a rod cell of a vertebrate photoreceptor, which is a modified neuron. A dense layer of photoreceptor cells at the back of the eye forms the retina, which is the structure that transduces light energy into neural information quanta. A single rod cell from the retina has an inner segment that contains the normal organelles of a cell and has a synaptic terminal at its base for transmitting information to other neurons. The outer segment of the rod cell is highly specialized and contains a stack of discs. These discs form at the base of the outer segment by invaginating and pinching off the cell membrane. The disc membranes are densely packed with rhodopsin and their function is to capture photons of light energy passing through the rod cell.

A rod cell converts a flash of light into a receptor potential through a chemical process whereby (1) a photon changes the form of a rhodopsin molecule which (2) activates a protein called transducin that (3) activates a phosphodiesterase which (4) controls the opening and closing of sodium ion channels in a membrane, (5) resulting in the flow of sodium ions that generate an action potential. This is a rather complex chemical amplification process where a single photon of light energy results in the closing of more than a million sodium channels, resulting in a change in the rod cell's receptor potential. Details of the chemical process, are provided by Purves *et al.* (1992).

From an information perspective, a photosensor converts a photon of light energy into neural information. The structures and methods for transmitting information from the photosensor to the iris, or to other light-responsive muscles, were discussed above in the section on structure.

8.5.5. Nonvolitional Behavior Related to Light Energy

The most obvious and most easily observed nonvolitional behavior in response to light energy (contraction of the pupil of the eye) is a means for automatically controlling the amount of light energy falling on the retina of the eye. This behavior allows us and other vertebrates to see relatively well under both high and low light intensities. The degree of this automatic pupil control is very fine—it is difficult, if not impossible, to observe the changes in pupil size as a function of very small changes in light intensity. This fineness of pupil control is due to the small sizes of the elemental behaving units and to the great number of units that are associated with pupil control behavior. The small behaving unit size and large number of behaving units also allow the pupil to maintain a particular size in response to a given intensity of light.

8.5.6. Quantification of Nonvolitional Light Behavior

The large number of animal species that have nonvolitional behavioral responses to light energy and the structural diversity of their light sensors complicates the process for quantifying these behavioral responses. Fortunately, light sensors are all of the same type and all depend on the rhodopsin molecule, as described above. The diversity in light behaviors among species is due largely to the ways the elemental light sensors and their associated muscles are structured and organized.

Once the structure and organization for a given individual are identified, the capacity to direct energy of the individual's behaving elements can be measured or calculated. The available energy required by the behaving elements can be quantified based on the elements' capacities to direct energy. The information generation and distribution structure of the behaving elements provide a means for quantifying the characteristics and amounts of information that will cause behaviors. The procedures developed earlier in this book can then be used to quantify the nonvolitional light behaviors for a particular individual.

Although calculations have not been made to illustrate the above-described processes, the treatment above is sufficient to validate the concept of nonvolitional light behavior depicted in Figure 8.3.

8.6. Mechanical Environment

Animals are subjected to mechanical forces in their environment, such as gravity forces, atmospheric pressures, wind, blows from other animals, and the like. These mechanical forces can be sensed by an animal's mechanosensors and can cause behaviors in response to these forces. The thrust here is the nonvolitional mechanical force behaviors. One readily observable nonvolitional mechanical force behavior is the "knee-jerk" reaction to a quick sharp pressure just below the kneecap. The knee-jerk behavioral response was used to test the validity of the quantitative mechanical force behavior concept described in the first part of this chapter and depicted in Figure 8.4. To demonstrate the validity of the mechanical force concept for a wide range of animals, the nonvolitional knee-jerk behavior and the simple sponge's behavior in response to mechanical forces were treated.

8.6.1. Structure

We consider the knee-jerk behavioral reaction first. The information generator is the muscle stretch sensor located in the muscle just below the kneecap. The distribution system is the nerves from the stretch sensor, to the dorsal root in the spinal cord, to the dorsal horn of gray matter in the spinal cord, to the motor neuron in the ventral horn of gray matter, to the ventral root, and then to the nerve to the appropriate behaving muscles. These muscles are the skeletal muscles that are involved in the knee-jerk behavior. A simple diagram of the cross-section of a human spinal cord showing the knee-jerk reflex arc is shown in Figure 8.5.

It was demonstrated by Parker (1918) in the early part of this century that the simple sponge has behavioral responses in its osculum (a small mouth) due to mechanical forces in the form of water flow. Parker reported that the closing of the osculum in quiet water and the opening in a current of water are both very local reactions and cannot be induced from points half a centimeter distant on the finger of a *Stylotella* sponge. Parker concluded from his research that sponges possess the beginnings of the neuromuscular mechanism but are devoid of true nervous elements. The sponge's structure consists of muscle tissue with some closely associated pressure sensitive tissue—the structure is devoid of differentiated sensors, information transmission, and muscle.

8.6.2. Capacity to Direct Energy

The methodology for determining the capacity to direct energy for the skeletal muscles involved in the knee-jerk behavior is, in essence, the same as that used for determining the capacity to direct energy of the sartorius motor unit of Rana pipiens. We first measure the total energy per contraction using the methods of Carlson et al. (1963).

Determination of the capacity to direct energy of the osculum closure behavior of a sponge is not as easy as it is for relatively large muscles such as those involved in the knee-jerk reaction. This is due in large part to the small sizes of the behaving elements involved in the osculum behavior and to the lack of a motor unit

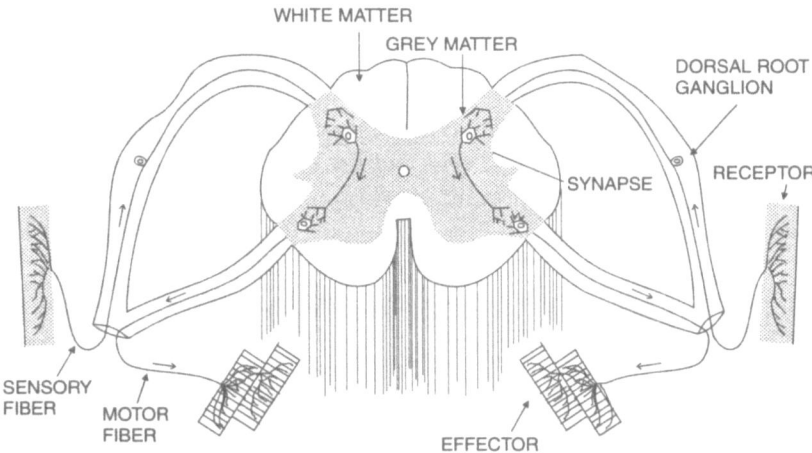

Figure 8.5. Nonvolitional reflex behavior.

as the behaving element. It is sufficient for our purposes to understand that it would be possible to measure the energy directed per contraction given the availability of instruments that could measure extremely small amounts of energy and given that the information generation rate of single molecules could be determined.

8.6.3. Available Energy

Determination of the capacity to direct energy for each contraction of a muscle behaving element provides a means for determining the amount of energy that must be available for each muscle contraction. The total amount of energy that must be available for each muscle contraction is equal to the total capacity to direct energy per contraction of the muscle behaving element.

8.6.4. Information Generation and Distribution

The information generators for the knee-jerk behavior are the stretch sensors in the tendons and muscles located in the front of the leg just below the kneecap. It is the tendon of the quadriceps muscle traveling over the knee joint that is tapped to elicit the knee-jerk reflex. This tap of the tendon activates the stretch receptors associated with this tendon. The stretch sensors consist of modified muscle fibers that are innervated in the center with sensory neurons. Whenever a muscle is stretched, the stretch sensor is activated, and the neurons of the stretch sensor transmit information to the central nervous system. The distribution of information from the stretch sensor to the central nervous system to the muscles involved in the knee-jerk behavior is described in the above section on structure.

The information generators for the opening and closing of the osculum are the surface cells that are sensitive to pressure. There is no true transmission system between these sensitive cells and the contractile cells. There is, however, a direct transmission of information from the generator to the contractile tissue.

8.6.5. Nonvolitional Mechanical Pressure Responses

The knee-jerk behavior is readily observed and is relatively straightforward. A sharp blow to the area just below the knee causes a stretching of the stretch sensors in this area, which generate information. This information is transmitted directly to the central nervous system where it is routed directly to the muscles associated with the knee-jerk behavior, causing these muscles to contract.

The behavior of the osculum is related to the flow of water over very specific areas of a sponge's finger. As such, the water flow rate is important because the pressure exerted on the sponge is related to the water flow rate. The osculum closes in quiet water and opens in a current of water.

8.6.6. Quantification of Nonvolitional Mechanical Pressure Behavior

The knee-jerk behavior is of special interest due to (1) its ease of observation, (2) the direct connection from the stretch sensor to the central nervous system to the responding muscle, and (3) the relative ease of measurement of information and behavioral energy. Items 1 and 2 have been discussed above, but the measurement of information and behavioral energy have not been covered.

Since the early days of elementary nervous system investigations, three important tools have been developed which allow measurement of information and behavioral energy. The scanning microscope provides a means for determining the minute structure of tissue and, thereby, a way to identify the number of motoneurons involved in the knee-jerk reaction. The number of motoneurons determines the number of motor units involved in a knee-jerk reaction. Microelectrode probes provide a means for both measuring electrical potentials in neurons and stimulating neurons. These probes provide a method for identifying the muscle fibers associated with a motoneuron. The processes of molecular chemistry provide a way to determine the chemical energy converted by muscle tissue, which allows the quantification of the total amount of energy converted by a motor unit. On a larger scale, the instruments for measuring mechanical force and energy have been improved so that the mechanical energy in the knee-jerk reaction can be measured.

The ability to isolate individual motor units and to determine the amount of energy used in response to a quantum of information provides a direct confirmation of the ability to quantify the knee-jerk behavior.

Due to the osculum's lack of differentiated sensors, neural transmission structures, and muscle tissue, that is, well-defined sensors and motor units, it is more difficult to determine the amount of chemical energy converted to work and to heat in response to information inputs. However, modern scanning microscopes and quantitative chemical procedures should allow for quantification of the amount of behavioral energy in response to a quantum information input.

The ability to quantify the knee-jerk behaviors as described above is sufficient to verify the mechanical behavior concept depicted earlier in Figure 8.4.

8.7. Summary and Conclusions

Nonvolitional behaviors are defined as those autonomous behaviors of animals that are caused by information generated as a function of environmental phenomena. The validity of the principles of a quantitative living systems science was tested using data on nonvolitional behaviors. The environmental phenomena used for these tests were temperature, gravity, light, and mechanical forces. A conceptual behavioral system was synthesized for each of these environmental phenomena. The principles were tested as they apply to each of the conceptual systems.

The following conclusions result from testing the principles of a quantitative living systems science using data on the nonvolitional behaviors of animals:

1. The easily observable nonvolitional behaviors of animals are readily described in terms of the principles of quantitative living systems science.
2. The easily observable nonvolitional behaviors of animals can be quantified using the principles of quantitative living systems science.
3. There is a very high confidence level in the validity of the principles as they apply to the easily observable nonvolitional behaviors of animals.

CHAPTER 9

Volitional Behavior

9.1. Introduction

The most obvious behaviors of animals are different from their autonomous and nonvolitional behaviors because they are rapid, variable, and easily observed motions. Our own rapid motions that can be readily observed by ourselves and others are those over which we have some control. The term *volitional behavior* is used here to mean those behaviors that are neither autonomous nor nonvolitional and over which we have the greatest control. Most obvious behaviors are the result of muscle contractions which are caused by coordinated information from the nervous system. An example of volitional behavior is a single deeper-than-normal breath caused by the integration and coordination of autonomous breathing information with information from our brain.

Behaviors caused by integrated and coordinated neural information are sometimes called conscious behaviors. However, conscious behaviors are considered by many to be a human characteristic not shared with other animals. Volitional behavior is a broader, non-prejudicial term that can apply to most animals. Volitional behaviors are caused by information that results from the integration and coordination of information from two or more information generators. This meaning of volitional behavior is sufficiently broad to cover the animal behaviors caused by the integration and coordination of neural information. Even so, volitional behavior is not a good choice of terms because it is used by some to designate only human behavior. Other terms have been used to describe the behaviors resulting from the combination of neural information. Early investigators, such as Parker (1918), have described this phenomenon in terms of adjuster (internuncial) neurones between a sensor and an effector. Miller (1978) uses the word *decider* for the melding of information and states that a decider is one of the characteristics of living systems at all levels of organization. With these caveats, the term volitional behavior is used to describe the subject of this chapter.

The easily observed volitional behaviors, such as standing, walking, and running, depend on the integration of neural information, which, in turn, depends on the structure and characteristics of neurons. Neurons are the cells directly responsible for sensing, processing, and integrating neural information and for generating commands to effector organs. Individual neurons make "decisions" about generating information (i.e., action potentials) by summing the information they receive from two or more sources. These neuron decisions cause the easily observed volitional behaviors.

There is a robust literature describing the structure and characteristics of neurons, the ways neurons are combined into nervous systems, and the processes by which information in nervous systems cause volitional behaviors. This descriptive literature was used as the basis for developing quantitative measures for volitional behaviors. This literature was also used for verifying the principles of quantitative living systems science as they apply to these observable volitional behaviors.

The characteristics and information processing behaviors of motoneurons are presented first because these neurons directly cause the readily observable volitional behaviors. An elemental behaving system consisting of a motoneuron with two information inputs is hypothesized for the purpose of illustrating the principles of information processing by a neuron. This hypothesized elemental system is used to demonstrate the quantification of information that causes volitional behaviors.

Unlike motoneurons, other neurons do not directly cause behaviors in muscles and organs. These neurons process information. The characteristics and behaviors of these neurons are presented after motoneurons. An elemental behavior system consisting of a neuron with two information inputs is hypothesized for the purpose of illustrating the principles of information processing by these neurons. We will demonstrate that the information generated by these neurons can be defined and quantified.

After validating the concept that the information generated by neurons can be quantified, the concepts of quantifying information and volitional behavior were tested and validated using data on food acquisition and thermal behaviors.

9.2. Conceptual Volitional Motor Unit System

The least complex system for combining neural information and causing volitional muscle contraction behaviors consists of (1) two neural information generators that transmit neural information to a single motoneuron; (2) a single motoneuron that combines the information inputs, generates new information from these inputs, and transmits this new information to the muscle fibers of a motor unit; and (3) motor unit muscle fibers. This conceptual system is shown in Figure 9.1. The information generators in this figure may be sensors or other neurons. The behaving motor unit of the conceptual system comprises the motoneuron and the muscle fibers.

The general characteristics and behaviors of this conceptual volitional behavioral system are determined from the easily observable behaviors and the characteristics of a motor unit. Like the easily observed autonomous and nonvolitional behaviors, easily observable volitional behaviors of animals are the result of motor unit contractions, and these contractions are caused by neural information. The neural information from the motoneuron is in the form of action potentials in the motoneuron's axon. These action potentials cause a motor unit's muscle fibers to contract and do work. Therefore, the only information output conditions from a motoneuron to the muscle fibers is either no output or an output that causes all the muscle fibers to perform work.

Observed volitional behaviors can be used to determine a motoneuron's general properties for combining two information inputs and for generating the neural information that causes the muscle fibers to do work. The standing and motion behaviors of vertebrates can occur because these animals have skeletal joints with paired muscles attached to opposite sides of the joints. Standing behavior results when the motor units of each muscle of the paired muscles are receiving information and the motor units are in a continual state of contraction, thereby holding the joints in positions associated with standing. The volitional behavioral system shown in Figure 9.1 represents one motor unit in one of the paired muscles. Motion behaviors of vertebrates result when one of the paired muscles of a skeletal joint contracts and the other muscle relaxes.

Determination of the conceptual system's characteristics and behaviors requires an understanding of the general structures, characteristics, processes, and behaviors of neurons.

9.2.1. Neurons

Neurons have three major types of neural structures as shown in Figure 9.2. They are (1) protoneurons (prototype neurons), (2) bipolar neurons, and (3) fully differentiated neurons. The protoneurons comprise the nerve-net of animals of

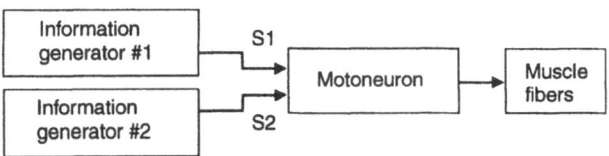

Figure 9.1. Conceptual volitional behavioral system.

low organization such as the sea anemone. These protoneurons do not have a synaptic system (Parker, 1918). The lack of a synaptic system precludes the integration of neural information and volitional behaviors. Bipolar neurons form the nervous system of slightly higher organized animals, such as the earthworm. At this slightly higher level of organization, the nervous system has become more differentiated and both polarity of information flow and synaptic relations have emerged. The early neurons of the neural structure of the earthworm provide a means for volitional behaviors. Fully differentiated neurons have dendrites, axons, and complete synapses. Vertebrates have fully differentiated neurons which provide a capability for complex information integration and volitional behaviors.

The properties of fully differentiated neurons are treated here because these neurons cause volitional behaviors. A typical differentiated neuron has a cell body and many extensions, as shown in Figure 9.3. Extensions called dendrites bring information into the cell body, and other extensions, called axons, conduct information away from the cell body. Information is passed between neurons and other cells at synapses where the membranes of two communicating cells come into proximity. The information transmitter is the presynaptic cell and the information receiver is the postsynaptic cell. There are two types of synapses. One is called electrical because its transmission process is electrical. The other is called chemical; its transmission process is chemical in nature.

An information quantum in a neuron is in the form of an action potential. An action potential is a sudden change in the resting electrical potential across a neuron's membrane. Resting and action potentials result from asymmetric distribution of ions on the two sides of the plasma membrane. An action potential is generated by selective opening and closing of specific ion channels which result in sodium ions moving into the axon, followed by potassium ions moving out. The

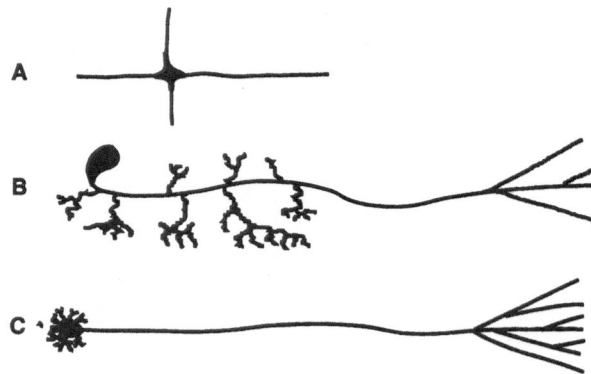

Figure 9.2. Neuron structures.

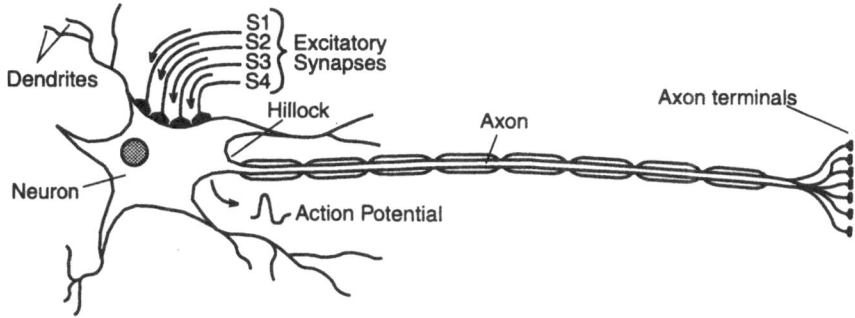

Figure 9.3. Differentiated neuron structure.

action potential travels down the axon of this presynaptic neuron to the axon terminals. At a chemical synapse, there are chemically gated ion channels in the postsynaptic cell's membrane that open in response to neurotransmitters. These changes in the ion channels can result in depolarizing or hyperpolarizing the postsynaptic cell's membrane and can make it more or less likely that the neuron will fire an action potential. Depolarizing synapses are excitatory, while hyperpolarizing synapses are inhibitory.

9.2.2. Muscle Fibers

The muscle fibers depicted in the conceptual system (Figure 9.1) are the postsynaptic cells that (1) receive neural information, (2) contract, and (3) perform the work that can be observed. The characteristics of these postsynaptic muscle cells are a determinant of volitional behavior and its quantification.

9.3. Motoneuron Characteristics

Quantification of volitional behavior requires an understanding of (1) the characteristics of a motoneuron, (2) how the motoneuron causes muscle contraction, (3) how a motoneuron receives information, and (4) how a motoneuron processes the received information to generate quanta of information that cause muscle contractions.

The characteristics of the information flow from the motoneuron to the muscle fibers are presented first because that flow is the direct connection between information and muscle behavior. In addition, the characteristics of muscle fibers place restrictions on the type of information the muscle will accept. The flow of information from the motoneuron to the muscle fibers is determined by the neuromuscular junction and the postsynaptic muscle membrane processes.

9.3.1. The Neuromuscular Junction

A motoneuron has axon branches with terminals that form neuromuscular junctions with individual muscle fibers of a motor unit. The axon terminals of the motoneuron (the presynaptic cell) come very close to, but do not touch, the muscle cells (the postsynaptic cells). A synaptic cleft of about 20 nanometers separates the membranes of the two cells.

An axon terminal contains many spherical vesicles filled with molecules called neurotransmitters. Motoneurons have neurotransmitters consisting of acetylcholine. Chemical synaptic information transmission begins with an action potential arriving at the axon terminal. The plasma membrane of the axon terminal has a type of voltage-gated ion channel found nowhere else on the axon: the voltage-gated calcium channel. The action potential causes the calcium channels to open. Because calcium ions (Ca^{2+}) are in greater concentration outside the cell than inside the cell, they rush in. The increase in Ca^{2+} ions inside the cell causes the vesicles full of acetylcholine to fuse with the presynaptic membrane and eject their contents into the synaptic cleft. The acetylcholine molecules diffuse across the cleft and bind to the receptors on the motor end plate causing the sodium channels to open briefly and depolarize the postsynaptic muscle cell membrane.

The postsynaptic membrane is part of the muscle cell's plasma membrane, which is slightly modified in the area of the synapse and is called a motor end plate. The modification that makes a patch of membrane a motor end plate is the presence of acetylcholine receptor molecules. The receptors function as chemically gated channels that allow both sodium (Na^+) and potassium (K^+) ions to pass through. Since the resting membrane is already fairly permeable to K^+ ions, the major change that occurs when these channels open is the movement of Na^+ ions into the cell. When a receptor binds acetylcholine, a channel opens and Na^+ ions move into the cell making the inside of the cell more positive.

9.3.2. Postsynaptic Membrane Process

The electrical properties of postsynaptic membranes are different from those of axon membranes in an important way. Because motor end plates have very few voltage-gated sodium channels, they do not fire action potentials. The binding of neurotransmitters to receptors and the opening of chemically gated ion channels perturb the resting potential of the postsynaptic membrane. This local change in membrane potential spreads to neighboring regions of the plasma membrane of the postsynaptic cell. Eventually, the spreading depolarization may reach an area of membrane that contains the voltage-gated channels. The entire plasma membrane of a skeletal muscle fiber, except for the motor end plates, has voltage-gated sodium channels. If a presynaptic axon terminal releases sufficient neurotransmitters to depolarize a motor end plate enough to bring its neighboring membrane to threshold, action potentials are fired. These action potentials are then conducted throughout the muscle fiber's system of membranes and cause it to contract. A large number of neurotransmitters are required to cause an action potential. Neither a single acetylcholine molecule nor the contents of an entire vesicle (about 10,000 acetylcholine molecules) is enough to bring the plasma membrane of a muscle cell to threshold. But a single action potential in an axon terminal releases about 100 vesicles, and that is enough to fire an action potential in the muscle fiber and cause it to contract.

9.3.3. Motoneuron Information Quanta

An important information characteristic of motor units is that their motoneuron action potential (information quantum) causes only excitatory postsynaptic potentials (EPSPs) in muscle fibers. This characteristic is due to the neuromuscular junction and the postsynaptic membrane processes described above. This characteristic is the basis for the all-or-none law of motor units. It is also the characteristic that provides the means for quantifying neural information as described in Chapter 4. That is, a quantum of neural information from a motoneuron causes the mechanical work performed during muscle contraction.

9.3.4. Motoneuron Characteristics and Information Processing

Motoneurons (1) receive information inputs from other neurons, (2) convert these information inputs to postsynaptic potentials, (3) attenuate these postsynaptic potentials as they travel from the synaptic site to the axon hillock, (4) combine these attenuated postsynaptic potentials at the axon hillock, and (5) generate an action potential, provided the combined postsynaptic potentials exceed the motoneuron's threshold voltage for generating an action potential. Figure 9.3 depicts a motoneuron with four information inputs that causes EPSPs at synapse sites S1, S2, S3, and S4 and an action potential at the axon hillock. Figure 9.4 depicts attenuated EPSPs from each of the four information inputs and shows the combination of EPSPs from three synaptic sites that exceed the action potential threshold. This figure also shows the action potential resulting from this combination of three information inputs.

A motoneuron may also generate postsynaptic potentials at its synapse with an input generator's axon terminals that inhibit the firing of an action potential by the motoneuron. These potentials are called inhibitory postsynaptic potentials or IPSPs. A brief summary of EPSPs and IPSPs provides a basis for quantifying volitional behaviors.

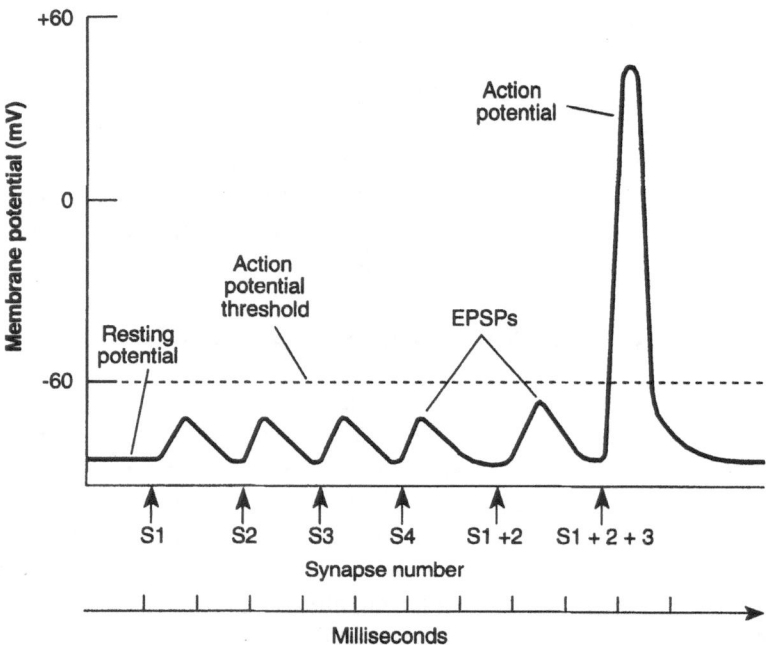

Figure 9.4. Spatial information summation.

Excitatory and inhibitory synapses are well-known phenomena. The axon terminals of presynaptic information generator neurons may make synapses with a motoneuron's dendrites and with its body. The axon terminals of different presynaptic neurons may store and release different neurotransmitters, and membranes of the dendrites and cell body of a postsynaptic motoneuron may have receptors to a variety of neurotransmitters. Thus a given postsynaptic motoneuron can receive a variety of chemical information. If the postsynaptic motoneuron's response to a neurotransmitter is depolarization, the synapse is excitatory; but if the response is hyperpolarization, the synapse is inhibitory. However, whether a synapse is excitatory or inhibitory depends not on the neurotransmitter but on the postsynaptic receptors—on what kind of ion channels the postsynaptic cell contains. The same neurotransmitter can be excitatory at some synapses and inhibitory at others.

The mechanization of inhibitory synapses is different in many respects from excitatory synapses. The postsynaptic cells in inhibitory synapses have chemically gated potassium or chloride channels as their receptors. When these channels are activated by binding with a neurotransmitter, they hyperpolarize the postsynaptic membrane. Thus, the release of neurotransmitters at an inhibitory synapse makes the postsynaptic cell less likely to fire an action potential.

There is a significant difference between the behaviors caused by information inputs to a motoneuron and the behaviors caused by its output. A quantum information (action potential) input to a motoneuron does not necessarily cause an information quantum output behavior of the motoneuron. By comparison, a quantum information input to a motor unit's muscle fibers causes them to contract. The information output behavior of a motoneuron is a function of the characteristics of (1) the axon terminals of the presynaptic neuron generating the input information to the motoneuron, (2) the type and number of neurotransmitters released by these axon terminals, (3) the synaptic structure of the motoneuron, (4) the location and distance of the synapse from the motoneuron's axon hillock, and (5) the properties of a motoneuron that cause attenuation of the potentials generated at synapses.

The presynaptic neuron's action potential causes chemical work to be performed to release these neurotransmitters. The amount of work performed to release neurotransmitters depends on the specific characteristics of the presynaptic neuron's structure. The neural information from the presynaptic neuron is carried to the postsynaptic motoneuron by these neurotransmitters. The type and number of neurotransmitters determine the properties of the neural information at this stage of transmission. The information in the neurotransmitters is converted back to bioelectrical form by the neurotransmitter receptor structures in the postsynaptic motoneuron.

The postsynaptic neural information in the form of bioelectrical potential spreads and is attenuated as it travels from the synapse site to the motoneuron's axon hillock. Therefore, this information is degraded and is less capable of performing work as it is transmitted from the synapse to the axon hillock. For example, an excitatory postsynaptic potential generated at a synapse near the axon hillock is more capable of performing work to generate an axon potential than a postsynaptic potential generated at a remote dendrite.

The information transmission processes just described are analogue phenomena as opposed to the on–off digital muscle contraction phenomena. This analogue phenomenon of adding electrical potentials provides a motoneuron with the capability to combine and integrate neural information from two or more information sources.

9.4. Volitional Motor Unit System Behaviors—Spatial Summation

The possible behaviors of the conceptual volitional behavior system shown in Figure 9.1 and the neural information that causes these behaviors are determined by the above-described characteristics of motoneurons. These possible behaviors are presented in Table 9.1.

The first column of Table 9.1 contains identifiers for the information condition at the input to the motoneuron and within the motoneuron. The second column gives the synapse associated with the motoneuron information input. The third column is the information input to the motoneuron—the input can be either a quantum of information (q) or zero. The fourth column is the condition where the sum of excitatory postsynaptic potentials is equal to or greater than the motoneuron's action potential threshold. Column 5 is the information condition where an inhibitory postsynaptic potential caused by information from one generator is sufficiently large that, when summed with an EPSP from the other generator, it results in a potential that is not large enough to cause an action potential. Column 6 is the condition where the sum of two EPSPs is less than the action potential threshold. Column 7 is the information condition where the EPSP is sufficiently greater than the IPSP that the sum of these two potentials is greater than the action potential threshold. Column 8 is the result of the motoneuron's information processing and is shown as either the presence or absence of an action potential.

Condition (a) in Table 9.1 is the lack of an input from either information generator. Therefore, there cannot be an excitatory postsynaptic potential that exceeds the action potential threshold of the motoneuron and there cannot be an action potential. This is the trivial condition of no information input to or output from the motoneuron.

Condition (b) is a quantum of information (q) from generator number 1 but no input from generator number 2. The addition of the potential caused by a quantum of information from generator number 1 and the lack of a potential from generator number 2 result in a potential at the axon hillock that exceeds the action potential threshold. Therefore, there is a quantum information output from the motoneuron.

Condition (c) is also a quantum of information from generator number 1 but no input from generator number 2. However, in this case the EPSP at the synapse caused by the input from generator number 1 has been attenuated to a value that is less than the action potential threshold by the time the EPSP reaches the axon hillock. Therefore, there is no information output from the motoneuron for this condition.

Condition (d) is a quantum input from each of the information generators. At the axon hillock, the sum of the EPSPs from these two inputs is greater than the action potential threshold. Therefore, there is a quantum information output from the motoneuron.

Condition (e) has the same inputs from the information generators as condition (d). However, at the axon hillock, the sum of the EPSP from these inputs is

TABLE 9.1. Spatial Information Summation

	Conditions	Input	Information Combinations				Action Potential
			EPSP>TH	IPSP>EPSP	EPSP<TH	IPSP>EPSP	
a	S1	0			0		
	+				+		0
	S2	0			0		
b	S1	q	P1				
	+		+				q
	S2	0	0				
c	S1	q			P1		
	+				+		0
	S2	0			0		
d	S1	q	P1				
	+		+				q
	S2	q	P2				
e	S1	q			P1		
	+				+		0
	S2	q			P2		
f	S1	0			0		
	+				+		0
	S2	q			P2		
g	S1	0	0				
	+		+				q
	S2	q	P2				
h	S1	q	P1				
	+						0
	S2	q		P2			
i	S1	q	P1				
	+						q
	S2	q				P2	
j	S1	q		P1			
	+						0
	S2	q	P2				
k	S1	q				P1	
	+						q
	S2	q	P2				

less than the action potential threshold. Therefore, the motoneuron does not have an information output.

Condition (f) is a quantum input from generator number 2 and no input from generator number 1. The EPSP at the axon hillock is less than the action potential threshold so the motoneuron does not have an information output to the muscle fibers.

Condition (g) has the same information inputs as condition (f). However, the sum of the inputs for condition (g) results in an EPSP that is greater than the action potential threshold.

Condition (h) is an EPSP caused by information from generator number 1 that exceeds the action potential threshold and an IPSP caused by information from generator number 2. The sum of these two potentials is less than the action potential threshold. Therefore, there is no action potential output from the motoneuron.

Condition (i) is an EPSP caused by information from generator number 1 that exceeds the action potential threshold and an IPSP caused by information from generator number 2. The EPSP is sufficiently greater than the IPSP that the sum of the EPSP and IPSP is a potential that is greater than the action potential threshold. Therefore, there is an action potential output from the motoneuron.

Condition (j) is similar to condition (h). The only difference is that the EPSP is caused by information from generator number 2 and the IPSP is caused by information from generator number 1.

Condition (k) is similar to condition (i). The only difference is that the EPSP is caused by information from generator number 2 and the IPSP is caused by information from generator number 1.

In addition to these 11 conditions, there is the trivial condition of an IPSP from each information generator. There cannot be an action potential from the summation of IPSPs.

Each of the information conditions listed in Table 9.1 and the behaviors associated with these conditions can be quantified for a specific motor unit. The work performed by the specific motor unit can be determined using the processes described in earlier chapters. The information from the motoneuron to the muscle fibers that causes this work can be determined for each of the information conditions.

The motoneuron depicted in Figure 9.3 has four excitatory synapses and therefore has more possible information combinations than the two synapses described above. One possible combination is shown in Figure 9.4. Other combinations of information inputs can have a form similar to those shown in the figure. The number of combinations can be determined and described using the procedures for the two information configurations presented above.

9.5. Volitional Motor Unit System Behavior— Temporal Summation

A motoneuron generates information not only from a combination of post-synaptic potentials from two or more synaptic locations but also from a combination of postsynaptic potentials from a single synaptic site that arrive at the axon hillock at approximately the same time. This phenomenon is called *temporal summation*. Figure 9.5 depicts the temporal summation of successive quanta of input information from synaptic site S1 shown in Figure 9.3. This phenomenon occurs when quanta of information from a single synapse site arrive at the motoneuron's axon hillock sufficiently close together in time that one quantum is influenced by following quanta.

The characteristics of motoneurons and of information provide the data necessary to determine temporal processing properties of the conceptual volitional behavior system shown in Figure 9.1. This determination is made by considering the possible conditions for sequential information inputs to a motoneuron and identifying the resulting states of postsynaptic potentials, axon hillock potentials, and information outputs.

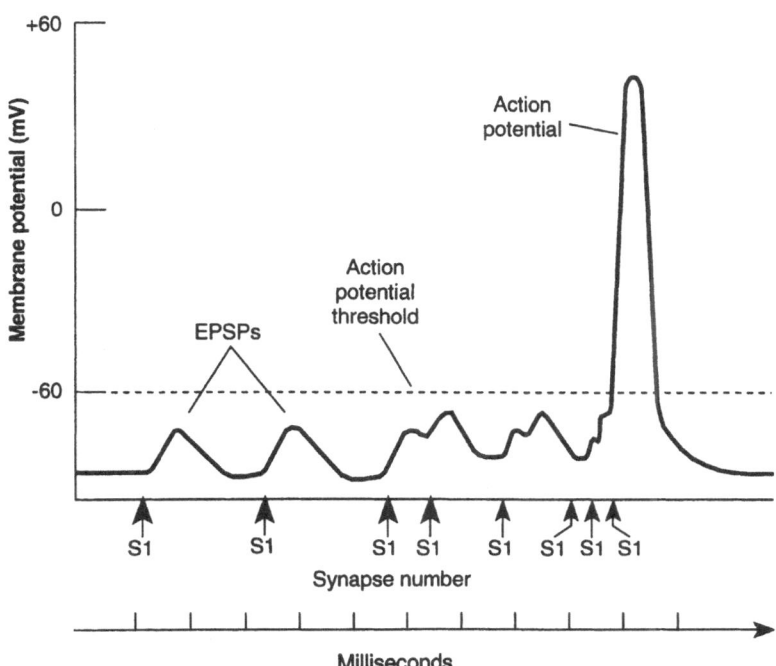

Figure 9.5. Temporal information summation.

An important characteristic of motoneurons in processing sequential information quanta from a single synapse is the time it takes for the motoneuron to return to its resting state after it has generated an action potential. The elapsed time from the initiation of an action potential until a motoneuron returns to its resting potential is approximately 2 milliseconds. After a motoneuron has recovered from generating an action potential, it can accept another EPSP at the axon hillock with no interaction between the first and the second information inputs. If the time between each quantum in a stream of quanta from a synapse is greater than a motoneuron's recovery time and each quantum generates an EPSP greater than the action potential threshold, each quantum of information input results in a quantum information (action potential) output from a motoneuron.

Other important characteristics of motoneurons in processing temporal information from one synapse are shown in Figure 9.5. This figure shows the processing of EPSPs at a motoneuron's axon hillock as a function of time. All these EPSPs have individual values that are below the motoneuron's action potential threshold. The two EPSPs on the left side of Figure 9.5 show that two sequential EPSPs separated by approximately 3 milliseconds are independent of each other and do not cause an action potential. The two EPSPs in the middle of Figure 9.5 show that two information inputs separated by approximately 1 millisecond do not cause an action potential even though the EPSPs are combined.

EPSPs that arrive at a motoneuron in rapid succession can cause an action potential. Figure 9.5 also depicts three EPSPs arriving at the axon hillock within approximately 1 millisecond causing an action potential. These three EPSPs arrive at the motoneuron's axon hillock in such rapid succession that these potentials add to provide a combined potential greater than the threshold potential. Although each excitatory postsynaptic potential is below threshold, the sum of the combined potentials is greater than threshold and causes an action potential.

The above temporal information processing conditions are summarized as follows:

- Each quantum of information in a sequence of quanta transmitted from a presynaptic information generator to a single synapse causes a motor unit contraction provided (1) each quantum causes an EPSP greater than the action potential threshold, and (2) the quanta in the sequence are separated by a time interval that allows recovery of the motoneuron and the muscle fibers.

- Two or more quanta of neural information in a sequence of quanta transmitted from a presynaptic information generator to a single synapse cause a motor unit contraction even when the EPSPs caused by each quantum are below the action potential threshold, provided the EPSPs at the axon hillock are sufficiently close together in time that the sum of these EPSPs is greater than the action potential threshold.

These information conditions and their resulting behaviors can be quantified by applying the processes described in previous chapters.

9.6. Neural "Decider" Conceptual System

There are two major differences between the structure and behavior (work done) of decider neurons and motoneurons. First, the information generated by a motoneuron causes mechanical work in the form of muscle contractions, whereas the information generated by decider neurons causes electrochemical work in postsynaptic neurons. Second, motoneurons cause only excitatory information in muscle tissue, whereas decider neurons can cause excitatory or inhibitory information in postsynaptic neurons.

Individual decision neurons make decisions about whether or not to fire action potentials by summing EPSPs and IPSPs. The ability of neurons to sum these potentials is the major mechanism by which the nervous system integrates information. Each neuron may receive 10,000 or more synaptic inputs, yet it has only one output—action potentials. All the information contained in the thousands of inputs a neuron receives is reduced to the rate at which that neuron generates action potentials. For most neurons, the critical area for decision-making is the axon hillock. The plasma membrane of the axon hillock is not insulated and has many voltage-gated channels that provide a mechanism for combining the EPSPs and IPSPs from anywhere on the dendrites or the cell body.

An action potential in a decision neuron's axon causes biochemical work to be performed in its axon terminals and in the postsynaptic neurons associated with these terminals. The information and behavior characteristics of decision neurons and their postsynaptic neurons are determined by considering a conceptual neural decider behavior system.

The least complex system for combining neural information that causes electrochemical work in a postsynaptic neuron consists of (1) two neural information generators that transmit quanta of neural information to a single decider neuron, and (2) a single decider neuron that combines the two information inputs, generates information, and transmits this resulting information to other neurons. This hypothetical conceptual system is depicted in Figure 9.6. The information generators in this figure may be sensors or other neurons.

At any moment in time, there are information conditions in this conceptual system that may or may not result in an action potential in the decider neuron, depending upon the specific characteristics of the neuron and the synapse.

Decider neurons have many of the same characteristics as motoneurons. A postsynaptic potential in both types of neurons is attenuated as the postsynaptic potential spreads from a synapse to the neuron's axon hillock, and excitatory and

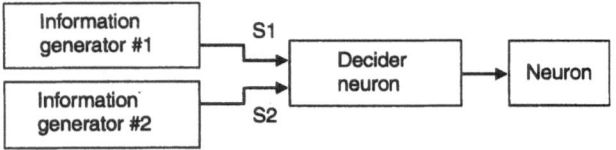

Figure 9.6. Conceptual decider behavior system.

inhibitory potentials combine the same way for both types of neurons. Therefore, the conditions for postsynaptic potentials in decider neurons are the same as those for motoneurons. However, the principal work caused by a motoneuron is the mechanical work of muscle contraction, whereas the work caused by a decider neuron is the bioelectrical and biochemical work associated with (1) the release of neurotransmitters, (2) the generation of postsynaptic potentials, (3) the overcoming of the resistance of the postsynaptic neuron to potentials, and (4) the generation of action potentials.

The characteristics of the system depicted in Figure 9.6 that determine the work performed by the information generated by a presynaptic decider neuron's action potential are listed below:

- The structure of the presynaptic decider neuron's axon terminal determines the type of neurotransmitters, the mechanisms for releasing these neurotransmitters, and the biochemical work required to release these neurotransmitters.
- The type of synapse and the structure of the postsynaptic neuron determine the work required (1) to accept the neurotransmitters, and (2) to generate EPSPs and IPSPs.
- The postsynaptic neuron's attenuation properties determine the work required for the transmission of postsynaptic potentials.
- The structure at the postsynaptic neuron's axon hillock determines the way potentials combine and cause the generation of action potentials in response to EPSPs and IPSPs.

The structure of axon terminals is complex and not completely understood. At present, about two dozen neurotransmitters have been discovered and possibly more will be found. The best understood are the acetylcholine neurotransmitters at all synapses between motoneurons and skeletal muscles, and the amino acid neurotransmitters that are prevalent in the brain's decider neurons. These amino acids are glutamic acid and aspartic acid, which are excitatory, and glycine and gamma-aminobutyric acid, which are inhibitory. Biochemical work is required for (1) synthesis of the neurotransmitters, (2) opening and closing of ion channels in membranes, (3) transport of the neurotransmitter-filled vesicles within the axon terminal, (4) fusion of the vesicles to the surface of the axon terminal, and (5) release of the neurotransmitters into the synaptic cleft. The energy to perform this work is supplied from biochemical energy sources such as ATP. However, documentation has not been found that quantifies the amount of energy and work associated with these processes.

The structures of the synaptic cleft and the postsynaptic neuron's membranes determine the amount of work necessary to transport the neurotransmitters across the synaptic cleft and to generate postsynaptic potentials. Diffusion of neurotransmitters across the synaptic cleft is caused by a difference in chemical concentrations. Work is necessary to cause these concentration differences. Binding of neurotransmitters to postsynaptic cells requires action by the neurotransmitter receptors of the postsynaptic cell's membrane. This process of binding requires the opening of receptor channels, which requires biochemical work. The generation of postsynaptic potentials also requires biochemical work. The values for the amount of work required to cause each of these processes has not yet been determined.

Reduction of postsynaptic potentials as they travel from a synapse to the axon hillock is caused by (1) spreading of the potential in many directions as it leaves the synapse and (2) attenuation within the postsynaptic cell. The energy in the postsynaptic potential is dissipated by both these processes. Spreading of the potential is a function of the cell's configuration and the spatial relationship of the synapse and the axon hillock. The potential attenuation between the synapse and the axon hillock is caused by the work required to overcome the cell's resistance to the flow of electrical energy. This work can be calculated provided the location of the axon hillock with respect to the synapse and the resistance of the cell are known.

Biochemical work is required to perform the processes of generating an action potential when the combined postsynaptic potentials exceed the cell's action potential threshold. Values for the amount of work required to cause action potentials remain to be determined.

The behaviors of the conceptual decider behavior system shown in Figure 9.6 are the combining of information and the generation of neural information. The work performed in the generation of information is the sum of the work from the processes presented above. It is theoretically possible to measure or calculate the work performed to carry out these processes. The end result of information is the work necessary to generate a quantum of information (i.e., an action potential).

9.7. Food Acquisition

The principles of quantitative living systems science presented in earlier chapters and the characteristics and behaviors of neurons presented above pro-

vide a basis for quantifying those volitional behaviors caused by neural information. These principles, characteristics, and behaviors were tested using existing
data for volitional food acquisition behaviors.

Following the approach of considering the most obvious and universal
behaviors first, the volitional behaviors of obtaining food were selected for testing the principles and information characteristics. The rationale for this selection
is that all animals must obtain food from their environment in order to live, and
they perform work to acquire food. The behaviors resulting from this work are
observable and animals with central nervous systems acquire food through volitional behaviors.

The volitional food acquisition behaviors considered here are the start of the
nutrition process—the sum of the processes by which an animal takes in and utilizes food. The remainder of the nutrition processes are autonomous and were
treated in Chapter 7 in the section on digestion. Volitional food acquisition behaviors include identifying food, performing those actions necessary to gather or
capture the food substances, processing the food to make it suitable for placing in
the mouth, processing the food in the mouth to make it suitable for entry into the
digestive structure, and causing the food substance to enter the digestive tract. As
an illustration, primitive humans search for substances that are identified as food,
gather or capture the food, prepare it for use (for example by removing the uneat-
able parts), place the food in the mouth where it is chewed, and swallow the
chewed food. Each of these behaviors is the result of the particular structure of
the animal, the utilization of muscle energy, and the generation of the information
necessary to cause muscle contraction. Also, all of these behaviors are volitional
as they are neither autonomous nor nonvolitional.

Following the approach used previously, food acquisition behaviors are considered in terms of the structures which provide a capacity to direct energy, the
available energy, the information which causes behavior, and the directed energy
in the behavior. As before, simple structures and behaviors are considered first,
followed by increasingly complex ones.

9.7.1. Food Acquisition Structures

To clearly differentiate the volitional food acquisition structures of multicellular animals from other nutrition structures, a short review of food acquisition
structures follows, starting from the lowest degree of complexity. These volitional behavior structures consist of skeletal systems, muscles, sensors, and information systems that cause muscle contractions. In contrast, single-cell animals have
membrane structures which provide a capability to absorb food into their bodies.
Indeed, each cell in an animal's body ultimately acquires its nutrients by absorption. At the next level of complexity, some single-cell animals have muscle struc-

tures, hydrostatic skeletons, and nervous structures which provide them with a capability to swim through nutrient-rich fluids from which they absorb nutrients. At this level, the muscle structure is cilia that move continuously in response to autonomous information. The sponge has a slightly more complex nutrition structure. It combines the cilia, which move nutrient-laden water through the structure of the sponge, with sphincter muscles that control this flow of water. The sea anemones are an example of animals with food acquisition structures with some increased complexity over those of the sponge. They have hydrostatic skeletons and muscle structures in the form of tentacles for obtaining food and for transporting this food to a mouth, which ingests the food, and an esophagus for transporting this ingested food into a digestive cavity. Their nerve structure is a nerve-net consisting of protoneurons.

The more complex animals follow this general structural plan of obtaining food and transporting it to a cavity or cavities where it can be digested. There are, however, many variations to this general structural plan which are a function of an animal's complexity. The more complex animals have exoskeletons and endoskeletons with their associated muscle structures and more elaborate nervous structures.

As a point of departure for considering nervous structures associated with the volitional acquisition of food, consider the nonvolitional food acquisition structures of actinians, specifically the sea anemones. According to Parker (1917), the feeding habits of sea anemones consist of operations that do not require the assumption of activities other than those consistent with the nature of a nerve-net. All the activities are local and the changes in behavior that they exhibit are due to fatigue. In these respects they are in contrast with the feeding habits in the higher animals that are controlled by a central nervous system. While almost all the elements involved in the feeding of actinians are isolated and act for themselves in a local way, hardly any such independence is observable in the elements concerned in the similar behaviors of higher animals. The jaws and their muscles, buccal glands, and so forth in these higher animals exhibit a highly unified action dependent chiefly upon the central nervous system. In contrast, the feeding habits in actinians emphasize the relative independence of parts rather than the action of the organism as a whole (Parker, 1918).

The least complex central nervous organs are found in worms. These animals do not possess the diffuse neural structure of the lower forms but, instead, have a centralized band of neural tissue extending the length of the body. In *Sigalion*, a marine worm, this band, which consists of a brain and a ventral nerve-chain, is part of the superficial ectoderm. In *Nereis*, the ventral chain is separate from the skin, though the brain is still a part of that layer. In the earthworm, the whole nervous band is in the deeper parts of the animal. This condition of a deep-seated central nervous organ is characteristic of the higher invertebrates such as worms, mollusks, and insects—just as it is for the vertebrates.

The receptor–decider–effector system described above provides the higher invertebrates and the vertebrates with a capacity for volitional behaviors. Receptor–decider–effector systems are presented below starting with simple structures and fundamental volitional behaviors, followed by more complex structures and behaviors of increasingly more complex animals.

9.7.1.1. Worms. Worms have a neural structure consisting of a superior nerve-ganglionic center (brain) which allows them to adjust their behaviors. The significance of this structure for more efficient food acquisition is indicated by the fact that a marine worm deprived operatively of its superior ganglion crawls about irregularly, does not turn readily upon contact with a corner, and burrows less readily than would a normal worm. Such a specimen, unlike the normal individual, is incapable of extending directly from its burrow and precisely seizing a shrimp or other passing organism. Removal of the worm's superior ganglion does not completely remove such normal reactions, although it does reduce considerably their directness and precision. At this stage of central nervous system emergence, the brain has a limited control of nervous impulses from the sensitive regions of the organism.

9.7.1.2. Insects. The central nervous structure of insects is more complex than that of the worm. The role of the brain of insects is that of an indispensable organizing center versus the superior transmitter role of the brain of worms. To illustrate this indispensability, upon removal of the head of a wasp or ant, it may walk about in a somewhat coordinated manner before dying; however, only scattered reflex components of normal behavior such as stinging may occur. The brains of insects, along with their muscle structures, are necessary for them to perform their complex food acquisition processes.

9.7.1.3. Octopuses. The structure of the octopus associated with food acquisition behaviors is more complex than those of worms and insects. It has a well-developed lens eye that has many individual receptor neurons. There are more motor units than in the less complex animals, and there are more deciders (i.e., central nervous system) neurons. This increased complexity resulting from more receptor, decider, and effector neurons provides increased capacity for control and precision in the behaviors associated with food acquisition.

9.7.1.4. Vertebrates. The behaving structures (sensors, deciders, motoneurons, and muscles) of cold-blooded vertebrates are, in general, more complex than those of invertebrates. This increased complexity is due, primarily, to a more elaborate tubular type of nervous system and an anterior brain.

The behaving structures of lower vertebrates provide a capacity for the basic vertebrate behaviors such as locomotion and the feeding lunge, and for a limited amount of volitional behavior modification through conditioning. The work of G. E. Coghill* demonstrated the relationship of these basic vertebrate activities to

---

* A comprehensive view of the development of behavior (including Coghill's work) is found in Leonard Carmichael, "The Onset and Prenatal Development of Behavior," *Manual of Child Psychology*, John Wiley and Son (1945)

the growth of the behaving structures of reflexive sensorimotor arcs in the nerve tube. Coghill's method in studying the appearance of fundamental activities in the salamander was to observe new changes in behavior and, for corresponding ages, to make parallel studies of the growth of nervous tissue. For example, Coghill traced the appearance of the complex S-movement of swimming in this amphibian to later and more complex growth changes in the nerve tube. The work shows that in the development of swimming in the salamander, the appearance of successive modifications in behavior depends upon the growth of new types of connective structures in the nervous system. Although locomotion and feeding lunge mechanisms are much more complex than those for lower invertebrates, such as the starfish, they are substantially on the same level of reflex-integrative behavior.

The behavior structures of the lower vertebrates make behavior modification possible through conditioning. In these animals, their structures are such that a large number of behaviors cannot be conditioned (i.e., they are nonvolitional behaviors). For example, experiments with the lizard *Lacerta* found distinct limits to the possibility of training where it opposed unlearned behaviors. Initially the animals were tested in their responses to simple geometrical figures which were moved in front of a black background. Their spontaneous lunges were most readily responsive to the movement of figures with even outlines such as circles, and least readily by figures with broken outlines such as triangles and crosses. In an attempt to change these behaviors, a training series was conducted in which the preferred circle was always presented with a quinine-treated mealworm mounted in front of it, the less-preferred cross with an untreated mealworm. In 850 trials no effective modification of the original responses appeared.

In submammalian vertebrates the behavior repertoire results mainly from their inherited structures and organization, with learning subordinate. This is true in the birds, which have robust genetically determined behaviors. In many of these behaviors, like migration, it is difficult to observe any essential contribution of learning in the principal behaviors. The case is different for pecking behaviors for which experience seems essential. Activity patterns, such as pecking in the domestic fowl, seem to depend upon a post-hatching process involving learning, in which certain early reflex-like behaviors are modified and integrated. Shortly after hatching, the component behaviors (head lunge, bill opening and closing, and swallowing) are essentially independent; however, through a gradual process of improvement, a fairly well-organized and accurate pecking response appears within four days. Chicks which are fed artificially and kept in the dark for a few days must go through the same process of gradual improvement, although in their case it occurs more rapidly in part because of organic developments such as greater muscular strength. Chicks which are artificially delayed longer than two weeks can be trained to peck at food only with great difficulty since they have learned to feed in other ways than by pecking. Therefore, maturation or growth dependent on hereditary factors does not contribute to the pattern or organization of pecking in the chick; rather, it contributes to the individual components (swallowing, etc.) which must be based on experience. Other evidence indicates that although the chick pecks at small bright objects without benefit of learning, it can discriminate between edible and inedible objects only on the basis of early experience in pecking.

The increase in complexity in the nervous structures of the lower vertebrates can be seen readily by comparing these structures with those of invertebrates. In the starfish, the nervous interconnections are few and the dominance of available centers is greatly limited so that very few adjustments can be made to the reflex behaviors. In the worms there is an advance in the sense that a principal nervous structure (center) has appeared, yet it functions largely as a superior transmitting center rather than an organizing center. The insect brain is superior in function, yet its organizing influence is rudimentary, as is shown by the fact that the behaviors which appear are always influenced by the characteristic reflexive properties of lower centers or local functions.

A similar condition prevails in the lower vertebrates. Comparative neurologists have shown that the brain of these vertebrates exercises only an incomplete control over the lower reflective systems. The brain of these lower vertebrates is not capable of discharging unitary patterns of impulses to lower centers under the influence of many sensory information inputs of a situation. The structure of the lower vertebrates which provides the capacity for more complex behaviors is a specialized type of tissue known as neopallium (new brain) or *cerebral cortex*. True cortex is present as a small laminated area in the upper forebrain of some reptiles and less clearly in birds. When the principal centers of the central nervous system are interconnected directly and indirectly through cerebral cortex, its presence permits the overriding of reflexive types of behavior by patterns learned through experience.

9.7.1.5. Mammals. The structures of mammals which provide them with their capacity for behavior are much more complex than those for non-mammals. These structures increase in complexity from marsupials to man and provide an ever-increasing capacity for volitional behaviors. This increased structural complexity is evidenced mainly by the number of neurons and synapses in the nervous system. As an example of this structural complexity, the human brain contains about 100 billion neurons, and each neuron may make synapses with a thousand or more other neurons—there may be as many as a million billion synapses in the human brain (Purves *et al.* 1992). Another indication of structural complexity is the optic nerve, which contains about one million axons (Purves *et al.*, 1992).

Although each neuron may receive 10,000 or more synaptic inputs, it has only one output—action potentials (Purves *et al.*, 1992). It is an action potential in a motoneuron that causes observable muscle behaviors. Therefore, it is the motoneurons that are the final integrators and processors of information which cause the contraction of skeletal muscles that is responsible for the readily observable volitional behaviors of animals. To illustrate the degree of complexity of skeletal muscle structures, humans have about 400 muscles on each side of their bodies—these muscles can consist of a number of motor units, and each motor unit may consist of up to 100 muscle fibers (Purves *et al.*, 1992). Many of these motor units may be utilized in the processes of acquiring food, preparing this food, and ingesting it into the autonomous digestive system.

9.7.2. Capacity to Direct Energy

The data provided earlier verifies the concept that an animal's volitional behavior is a function of its muscular and skeletal structures, and of a central nervous system which can adjust the information input to skeletal muscles for various environmental and internal conditions. It is seen that an animal's capacity to direct energy is a direct function of its muscle and skeletal structures. Previously, the energy in a motor unit was determined for specific muscles (e.g., the Rana pipiens sartorius muscle). However, a more general determination of the energy utilized in a single muscle contraction can be made. A single muscle contraction of given strength in a given length of muscle is always associated with a fairly constant utilization of total energy. If T dynes is the tension developed in the isometric contraction of a muscle fiber l cm long, and if H ergs is the total energy utilized, the relationship among these variables for all muscles is approximately $H/Tl = 1/3$.*

As shown previously, this energy is the result of the muscle converting chemical energy into work and heat in response to an information input which causes contraction of the muscle. This equation shows that muscle has approximately a 30% efficiency in converting chemical energy into mechanical work. This compares to the 20%–28% efficiency described for the human heart in the Appendix.

9.7.3 Information Generation and Processing

The volitional-food gathering behaviors of animals have common information generation and processing characteristics. An animal's sensors of its internal conditions generate information on the need of the animal's cells for nutrition. This information is transmitted to the neurons associated with food acquisition,

*These data and the equation were obtained from *Encyclopaedia Britannica*, Vol. 15, 1951, p. 983.

such as the food-detecting sensors. Sensors of the external environment generate information on the presence and location of food and transmit this information to neurons associated with the skeletal muscles that cause equilibrium and motion behaviors. This information on food presence and location is combined with the information from the gravity and muscle stretch sensors that cause equilibrium and motion behaviors. This combined information causes the coordinated skeletal muscle contractions that result in food acquisition behaviors.

The common information generation and processing characteristics are expressed in as many forms and complexities as there are animal species with volitional behaviors. These forms and complexities range from the simple sensor and neural systems of worms to the robust sensors and brain of the human species.

9.7.3.1. Worms. Worms have few and simple chemical and tactile sensors, their nerve ganglionic center has relatively few decider neurons, and they have a relatively small number of skeletal motor units. Chemical sensors monitor the chemical process of an animal's cells and generate information based on the chemical conditions. These chemical sensors generate information on the need for food just as the internal chemical sensors described in Chapter 7 monitor the need for oxygen. A worm detects and locates food in the environment with its chemical sensors by sensing chemicals emanating from the food. These chemical sensors generate information in response to food and transmit this information to decider neurons in the ganglionic center. There it combines with information on the need for food. If there is no need for food, the decider neurons do not generate additional information. If there is a need for food, the decider neurons generate information and transmit this information to the neurons that cause the equilibrium and motion behaviors. There, this information is combined with the information generated by muscle stretch sensors and gravity sensors and generates the information that is transmitted to motoneurons. This information generation and processing causes a worm's food-locating and food-approach behaviors.

The information generation and processing operations for food detection are more complex than described above. The neural system processes information from chemical sensors and decides whether or not the chemicals sensed are associated with food. These operations are determined by the specific structure and organization of the species. For example, the chemical sensors may be so specific that they will only sense the chemicals associated with food, or the brain neurons may be structured by genetics to generate information only in response to specific chemical information generated by chemical sensors. More complex species may identify food by the brain's decision neurons, learning which chemical information represents food.

Ingestion of food involves the generation and processing of tactile information. A worm species with only chemical and tactile sensors for food acquisition generates information about the proximity of food through these sensors. Chemical sensors provide information on the direction to food, and tactile sensors provide information on contact with food. Tactile information is transmitted to the decision neurons that cause (1) the volitional opening and closing of the worm's mouth, and (2) volitional food processing prior to the food entering the autonomous segments of the digestive system. Tactile information is combined in these decision neurons with information from the muscle stretch sensors and, where applicable, with gravity sensors. The information generated by these neurons is transmitted to appropriate motoneurons. In general, there is a continuous flow of information to these motoneurons to maintain the muscle contractions associated with autonomous behaviors. The volitional food acquisition information from the decider neurons is combined in these motoneurons with the information that causes autonomous behaviors. The information generated by these motoneurons, combining the volitional and autonomous information, is transmitted to the appropriate muscle fibers to cause the muscle contractions that are the final observed behaviors of food acquisition.

The low complexity of a worm's information generation and processing systems allows these systems to be described accurately. Current instruments, research techniques, and the existing results of prior research provide a basis for the construction of a representative descriptive model of these systems. An accurate descriptive model of information generation and processing systems, along with the information processing principles presented earlier in this chapter, can be used to quantify the information generation and processing of the food acquisition behaviors of worms.

Some worm species have simple visual sensors that generate information useful in locating and acquiring food. This information is transmitted to decision neurons in the central neural system where it is combined with chemical information to enhance the information used in food acquisition behaviors.

The information generation and processing systems of worms are not static. These systems change in structure and capability as a worm develops from a single cell to maturity. The sensor's structure develops and the number of neurons and their connections increase as the worm grows. In addition, the central nervous system's information processing structure changes in response to changes in the environment.

9.7.3.2. Insects. Information generation and processing by insects is much more robust than that for worms because insects have many more information generators and an increased information processing capability. This increased robustness is due in part to an increased number and complexity of sensors, such as eyes, and a greater number of decision neurons that generate information. These enhanced information generation and processing characteristics provide insects with a capability for more robust autonomous, nonvolitional, and volitional behaviors than those of worms.

An insect's eyes generate information that allows identification and location of food at a distance. Visual information about food at a distance is generated more rapidly than chemical information due to the large difference between the speed of light and the speed of chemical diffusion. This rapid location of food and the increased mobility of insects with respect to worms result in the quicker food acquisition behaviors of insects.

Information generated by the eyes is transmitted to specific decider neurons in the brain where it is processed to identify food. Then, (1) information generated by the food identification and location neurons is combined with information generated by the internal sensors that sense the need for food, (2) information resulting from this combination is combined with the equilibrium and motion information generated by gravity and muscle stretch sensors, and (3) the information generated by this latter combination is transmitted to appropriate muscle fibers to cause the coordinated muscle behaviors associated with movement to the food source.

Insects with audio sensors can also detect and locate food sources with information generated by these sensors. The general plan for the generation and processing of information from audio sensors is similar to the information generated by visual sensors.

The visual and audio sensors of insects also generate information that can be combined with the information generated by tactile sensors to augment food ingestion behaviors. Visual and audio information can be processed and used to cause the mechanical motions of capturing food, bringing it to the mouth, and placing food into the mouth.

9.7.3.3. Octopuses. The information generation and processing systems of the octopus are more robust than those of insects because of the large number of axons in the octopus' eyes and in its central nervous system. All of the axons in the eyes are information generators that transmit information to decision neurons in an octopus' central nervous system. These decision neurons generate information that may be transmitted to other decision neurons for combining with information from other sensors (e.g., gravity and stretch sensors) and with autonomous and nonvolitional information. Because of the large number of neurons, synapses, and motor units, it is difficult to precisely identify and define the individual information generators and the processing of the information from these generators.

9.7.3.4. Vertebrates. Vertebrates have complex and sophisticated information generating and processing systems to complement their more complex and capable sensors and effectors. These information generating and processing systems are distributed among sensors, a sophisticated central nervous system, and various effectors. These attributes provide vertebrates with a capability for a number of volitional equilibrium and motion behaviors that enhance volitional food acquisition behaviors. The muscles associated with these volitional behaviors are many and range in size from a few to a relatively large number of motor units. The amount of information generated by motoneurons to cause these volitional food acquisition behaviors is also large and varied. Vertebrates have a large number of information generators in their sensors and central nervous systems. A large number of elemental information generation and processing systems like those shown earlier in the chapter is necessary to generate the information from multiple sensors, process this information, and cause the coordinated contractions of muscles.

Because of the large number of neurons and synapses in the information generating and processing systems of vertebrates, it is very difficult to precisely identify and specify the structure and nature of these systems.

9.7.3.5. Mammals. The great complexity of mammals' information generation and processing systems is illustrated by the approximate billion neurons in the human brain—each neuron is an information generator and processor. The number of volitional information generator and processor neurons is only a fraction of the brain's total number of neurons. Many of the brain's neurons cause autonomous and nonvolitional behaviors. The volitional information generators and processors in the brain are those used to generate the information that is ultimately transmitted to the motoneurons that cause volitional muscle contractions. The number of motor units and their motoneurons is believed to be relatively small compared to the number of decider neurons that generate information which can influence motoneurons.

9.7.4. Food Acquisition Behaviors

Volitional food acquisition behaviors are determined by the structures and information generation and processing systems presented above. A description of these behaviors is given below in terms of these determinants. The description proceeds from animals with the least complex volitional food acquisition behaviors (worms) to those with the most complex volitional food acquisition behaviors (mammals). The behaviors described include readily observed behaviors such as placing food into the esophagus and the difficult-to-observe behaviors, such as detection of food.

9.7.4.1. Worms. Obvious and directly observable volitional food acquisition behaviors of worms include sensing food, capturing the food, and transporting it to the mouth. Because the sea worm is mobile, coordination is required between muscles used for motility and those used to capture food and bring it to the mouth. Motility requires the establishment of a fixed reference with respect to the gravity gradient, and food gathering requires motion with respect to this reference. Therefore, there must be coordination between the muscles that maintain the worm "right side up" and the muscles that capture and transport food to the mouth. This coordination is achieved through the central nervous system by its integration of information from the sensors of gravity and of food location; and by its generation of information which is transmitted to the muscles for keeping the animal oriented with respect to gravity and the muscles involved in the capture and transport of food to the mouth. This integration process by the central nervous system combines nonvolitional behaviors, such as orientation with respect to the gravity gradient, with the volitional behaviors of acquiring food.

Learned modifications to volitional food gathering behaviors have been achieved in worms using food stimuli. For example, the marine worm *Nereis* has been trained to appear from a glass tube when it was lighted or when it was darkened depending on which of these visual changes had been combined with meat juice during the training period. At the initiation of training, light was not effective in causing the worm to appear from its tube. Meat juice, however, was effective, and the application together of these two stimuli produced changed organization whereby light possessed a new type of control over behavior. The ability to learn from the environment and to adjust behaviors is a major characteristic of volitional behaviors.

9.7.4.2. Insects. The volitional food gathering behaviors of insects are more complex than those of worms due, in large part, to their increased motility and to their visual sensors. For example, some insects have wings and can fly. Flying requires a high degree of coordination of the muscles associated with flight. This coordination is provided by a central nervous system that is capable of rapidly processing information from a number of information generators. In general, insects acquire food by going to sources of food as opposed to remaining in one place and gathering food as it goes by. There are so many species of insects that it is difficult to generalize their food acquisition behaviors. For example, insects have many means of locomotion, including flying, walking, crawling, swimming, and burrowing. However, they do have a number of food acquisition behaviors in common. They identify sources of food, travel to and return to sources of food, gather food in some manner, and ingest it. The identification of food by the eyes provides a capability which is versatile, can be used over relatively long distances, and has a rapid information generating capability. To process this greater volume and speed of information, the central nervous system must have a higher capacity than that of the less complex animals with their less capable sensors. A more capable central nervous system is also required for the coordination associated with the complex behaviors of insects.

9.7.4.3. Octopuses. The volitional food acquisition behaviors of the octopus are complex due to the capability of its eyes, to its more sophisticated central nervous structures, and to its numerous motor units. The octopus can make adjustments to changes in the food environment insects cannot because the octopus has greater capability to modify its nonvolitional behaviors than do insects.

The increased capability of the octopus' structure and information processing system provides improvements in the precision and variation of orientation. For example, the octopus moves forward only when an object, such as a crab, appears within a given range of distance. Numerous experiments indicate that it is the specific quantitative characteristics of stimulation (e.g., intensity, size, amount of change, rate of movement) that control the timing and precision of approach and withdrawal. Simple factors such as these appear to control the behaviors of the octopus. The fact that it has a well-developed lens eye approaching that of many mammals in complexity does not mean that it "recognizes" its food as do higher mammals. For example, when an octopus was presented with a crab in full view within a glass jar, the crab was seized only when a tentacle just happened to come into contact with it, without any directive aid through vision.

9.7.4.4. Vertebrates. The volitional food acquisition behaviors of vertebrates are both diverse and complex. They may have food sensors that span the full range of observable physical phenomena. Some animals sense electrical emissions from their prey, most sense chemical aerosols given off by prey, some sense acoustical emissions from prey, and a large number of them sense visual phenomena associated with their food. The ways they capture and gather their food is also diverse and is a function of their skeleton and of the skeletal muscles. But most of all, the enhanced behaviors of vertebrates are observed to be related to their central nervous system. Their behavioral capability to adjust to environmental conditions is observed to increase with increases in the size of the brain, that is, in relation to the number of information generators that can be used to "adjust" their behaviors. This increase in versatility is illustrated by the progressively complex food acquisition behaviors of vertebrates as their central nervous structures, in particular their brains, contain more and more neurons.

9.7.4.5. Mammals. Mammals exhibit great diversity in their food acquisition behaviors. These behaviors range from those of lower mammals, which are like nonvolitional behaviors, to those of humans, which have the greatest variety of all animals. Some mammals have discovered how to acquire and use tools in the acquisition of food. The diversity of food acquisition behaviors of mammals is the result of a mammal's capacity to learn differences among environmental conditions. For example, the object or incentive which attracts a hungry mammal tends to be discriminated from others on the basis of stimulus properties governed largely by individual past experience. The particular individual food habits of many pet dogs and cats illustrate the point. There is also an important difference between lower and higher mammals which comes into play in primates and especially man: the ability to have a food need satisfied by objects or conditions only indirectly related to food. A pat on the head stimulates a trained dog to further efforts, but a rat can acquire such special learned control only slightly if at all. Chimpanzees may be trained to work for tokens which later may be inserted into a food-delivery slot, thus the inedible object itself acquires a measure of food drive satisfaction for the animal. This is a capacity for symbolic satisfaction which reaches its greatest development in the highly indirect power of words over human behavior.

An animal's capacity to learn differences among environmental conditions depends first on sensory discrimination—learning to distinguish and behave appropriately according to simple sensory differences. Mammals are less tied to unlearned responses and are better learners than lower vertebrates and therefore excel in this capacity. Dogs learn innumerable fine odor discriminations; monkeys learn many delicate visual discriminations. In a test of weight discrimination in monkeys, the animals soon learned to pull in by its string the heavier of two visually identical boxes. From our point of view, a mammal's increased capacity to learn differences among environmental conditions and thereby have more complex food acquisition behaviors, is a function of increased sensory differences resulting from an increased number of information generations which can respond to environmental conditions.

Learned differences among environmental conditions depend secondly on the mastery of a *schema*, or pattern of sensory cues, to the extent that giving a response may depend on special conditions such as the presence or absence of certain parts of the pattern. For example, at first a young hunting dog may turn to follow almost any animal scent; later, if properly trained, he will track only pheasant and only if the scent is fresh. This type of accomplishment, which has been called *conditional discrimination*, places far greater demands upon learning ability than does simple discrimination and, unless very simple, is beyond the capabilities of lower mammals. From a volitional behavior perspective, the information distribution from the environmental sensor information generators is through a number of synapses which establish preferred pathways that represent a learned schema. The capacity for mastery of schema is therefore a function of both the number and characteristics of synapses.

Appropriate response to the perceptual schema becomes more difficult when it is encountered in a variety of situations, accompanied by variations which initially disturb the animal. A control over the essential significance of the schema, despite the presence of various confusing meanings under different conditions, marks the mastery of a more complex set of relationships frequently termed *abstractions*.

Discerning a given pattern in varying circumstances, amid confusing differences and partially similar features, is an accomplishment in learning relationships of which only the higher mammals are clearly capable. The limit of the lower mammals for "abstraction" is soon reached. In contrast (as N. Kohts has shown with her matching-from-sample technique), the chimpanzee is far more versatile and can single out the given schema (e.g., triangle), although it is presented in different sizes and colors and among other figures similar to it in color and size. The key for perception is not the presence or absence of given features of sensitivity (such as color vision), but rather, the given animal's ability to organize sensory data in dealing appropriately with new situations. This is critically dependent on the animal's capacity for learning new adjustments which, in turn, depend on a robust central nervous system.

This capacity for learning new adjustments provides mammals their wide diversity of food acquisition behaviors. There is a significant correlation between mammals' food acquisition behaviors and the number of environmental sensor-information generators and the number of synapse-information generators. That is, there is a significant correlation between mammals' central nervous system size and complexity and the variety and complexity of their food acquisition behaviors.

9.7.5. Quantification of Volitional Food Acquisition Behaviors

The above data describe (1) the capacity to direct energy, available energy, and information determinants of volitional food acquisition behaviors and (2) the relationship among these determinants. These data provide a qualitative verification of the behavioral relationship and principles developed early in this book as they apply to volitional food acquisition behaviors. These data also provide the basis for the quantification of volitional food acquisition behaviors and for further confirmation of the principles of quantitative living systems science.

Quantification of volitional food acquisition and ingestion behaviors is considered first because all animals must ingest food. These behaviors are the volitional food processes just prior to food entering the autonomous digestive structure. The breaking up of food into sizes that can enter the esophagus is an example of this behavior. The structures for these volitional food processing behaviors are species-specific; however, they all have muscles that cause these observable behaviors. The energy in these muscle behaviors can be described, and the energies utilized in the behaviors can be either measured or otherwise quantified. The motoneurons associated with the muscle contractions can be identified and the number of quanta associated with the behaviors can be determined. In simple animals with primitive central nervous systems, it is possible to describe the neural circuits consisting of information generators associated with the body's need for food and decider nerves such as those associated with the information generators indicating the presence of food in the mouth. Therefore, it is possible to determine the information processing associated with these simple behaviors. Even in complex animals, such as the monkey, the nerves associated with the contraction of skeletal muscles have been traced using microelectric probes to stimulate the central nervous system and then observing the resulting muscle contraction. Hence, it is possible to determine all the elements necessary to quantify behavior, that is, the capacity to direct energy, the energy in the behavior, the available energy, and the information associated with the behavior. To elaborate, (1) a specific animal's capacity to direct energy can be determined from the structures of its sensors, nervous distribution system, skeleton, and skeletal muscles, (2) the energy in the behavior can be determined by direct measurement or by calculation, and (3) the information associated with the behavior can be determined by direct measurement. The measurement of information quanta in nerves is possible through rather recently invented measuring devices, such as magnetic resonance instruments and very sensitive microelectric probes.

Data are presented above for the volitional food acquisition behaviors of taking food from the environment and placing it in the mouth. These behaviors are more complex than the ones just described for food ingestion because more sophisticated sensors of the environment are required to identify food, locate the food in the environment, and capture the food. Also, more sophisticated muscles and their motion with respect to a gravity reference are utilized in these behaviors, and a more complex central nervous system is necessary. However, the basic processes are the same for determining the animal's capacity to direct energy, the energy utilized in the behavior, the available energy, and the information which causes the behavior. Although the processes for quantifying these behaviors are more involved than the simpler behaviors described earlier, they can be used to quantify these volitional behaviors of capturing food and placing it in the mouth.

The basic processes and procedures just described for quantifying the simple volitional food acquisition behaviors are, with some elaboration, applicable to the more sophisticated behaviors associated with complex animals and their behaviors. This elaboration is necessary due to the more diverse behaviors of these animals, the greater complexity of the information generators, and, especially, the structure and number of synapses of the central nervous system.

9.8. Thermal Phenomena Volitional Behavior

All animals have behaviors which allow them to cope with the ubiquitous and ever-changing thermal environment. An animal's very survival depends on the behaviors which allow them to survive in thermal environments. Some animals have autonomous behaviors which allow them to cope with thermal environments. Others use volitional behaviors to survive and function in these environments. The most universal and obvious volitional thermal behaviors are treated here. These behaviors are the physical movements from one local thermal environment to another in order to obtain the desired body temperature.

Animals exhibit volitional thermal behaviors when they alternate between sunny and shady places in order to maintain their body temperatures within a given range. Ectotherms such as lizards regulate their body temperatures by volitional behavior. In a desert environment, the lizard maintains its body temperature by basking to absorb solar radiation in the cool mornings, whereas at midday it avoids the sun and maximizes its heat loss to the cooler air. This behavior allows the lizard to maintain its body temperature during the day at about 35°C.

It was hypothesized that these volitional thermal behaviors can be quantified using the concepts and principles developed herein. This hypothesis was tested using available data on thermal behavior phenomena.

9.8.1. Structures

The basic structure associated with the behavior of moving from one local thermal environment to another consists of (1) the skeletal and muscular structures for equilibrium and locomotion; (2) sensors for determining the thermal environment; (3) an internal information generator for establishing a "thermal set point" which determines the optimal internal temperature of the animal; (4) a central nervous system that can determine the difference between the external thermal environment and the "thermal set point," learn the environmental characteristics that represent warm and cold conditions, and generate adjustment information; and (5) a distribution system that transmits information to motoneurons. The information transmitted to motoneurons results from the integration of information generated by the thermal environment sensors, the thermal set point information generator, and the information generated by the brain's memory structure. Data on the structures of the ectotherms, such as the lizards, validate this basic conceptual structure.

The purpose of choosing the ectotherms is that they depend on external sources of heat, such as solar radiation, to maintain their body temperature above the environmental temperature, and they have behavioral mechanisms for maintaining optimal body temperatures. Within the ectotherms, the desert lizard was selected due to the large temperature differences it must cope with in its desert environment. The temperature environment in the desert can change more than 40°C in a few hours.

Desert lizards have skeletal structures and skeletal muscles that provide them a capability for rapid and sustained locomotion. They have temperature sensors in their skin which generate information related to the heat condition of their environment. The structure of their brain includes a hypothalamus which provides the "thermal set point," that is, the optimum temperature information. Lizards have sensors, such as eyes, a central nervous system, and a brain which provide an adequate structure for learning about their environments. These structures provide a capacity for (1) learning which environmental characteristics and features represent warm and cool environments, and (2) generating the information necessary for locomotion to an appropriate thermal environment.

9.8.2. Capacity to Direct Energy

The capacity of an ectotherm to direct energy for thermal behaviors is a direct function of its structures, just as the capacity to direct energy for other behaviors is a direct function of structure. The relationships between an animal's capacity to direct energy and structures have been treated earlier for various behaviors. The same relationships hold here.

9.8.3. Available Energy

The chemical energy available for conversion to muscle and heat energies is, in principle, the same for all muscle behavior. However, for ectotherms there are subtle differences in the energies available for thermal volitional behaviors. The availability of chemical energy for use in muscles is a function of the metabolic processes of an animal. But, for ectotherms, the metabolic process is a function of the temperature of the animal. For low body temperatures, the metabolic processes slow down and for high body temperatures, the metabolic processes speed up. Therefore, at low temperatures an ectotherm moves slowly, and at high temperatures it can move much more rapidly.

9.8.4. Neural Information Generation and Distribution

The neural information generation and distribution systems associated with volitional thermal behaviors cause movement from one thermal environment to another. These systems interact with the information generation and distribution systems that cause equilibrium and movement behaviors.

Desert lizards have complex neural information generation and distribution systems for volitional thermal behaviors compared to lower ectotherm forms. Desert lizards have highly developed sensors and central nervous systems that can distinguish complex environments, generate schema from the sensor information, create abstractions of the environment, learn appropriate behaviors, and cause the muscle contractions associated with these behaviors. Lower ectotherms may have systems that generate information by sensors that only detect thermal gradients and distribute this information to simple muscle effectors that result in rudimentary swimming behaviors.

The desert lizard has sensors that are used to distinguish temperatures at their skin and eyes that can sense sunny and shady environments. They have a central nervous system that can combine information from their sensors with internal thermal set point information and with information generated in their brain that provides a representation of the thermal environment. The resulting information from this combination is transmitted to the various motoneurons that cause equilibrium and motion behaviors. The coordinated information transmitted to the appropriate motoneurons causes the observed volitional thermal behaviors.

9.8.5. Thermoregulatory Behaviors

The desert lizard has obvious, yet complex, thermoregulatory behaviors. There are actually two temperature "set points" for the desert lizard. At night, when the temperature of the desert may drop close to freezing, the temperature of the lizard remains stable at 16°C. This is achieved by the lizard spending the night in a burrow where the soil temperature is a constant 16°C. Early in the morning, soon after sunrise, the lizard emerges from its burrow. The air temperature is still quite cool, but the body temperature of the lizard rises to 35°C in less than 30 minutes. The lizard achieves this by basking on a rock with maximum exposure to the sun. As its dark skin absorbs solar radiation, its body temperature rises considerably above the surrounding air temperature. By altering its exposure to the sun, the lizard maintains its body temperature at around 35°C all morning as it seeks food, avoids predators, and interacts with potential mates or competitors. By noon the air temperature near the surface of the desert has risen to 50°C but the lizard's body temperature remains around 35°C. It is now staying mostly in shade, frequently up in bushes where there is a cooling breeze. As afternoon progresses, air temperature declines, and the lizard again spends more of its time in the sun and on hot rocks so that its body temperature still remains around 35°C. It returns to its burrow just before sunset, and its body temperature rapidly drops to 16°C. This demonstrates that the lizard can regulate its body temperature quite well by behavioral mechanisms rather than by metabolic mechanisms. A very interesting thermal phenomenon is that if a lizard is infected with pathogenic bacteria, it will give itself a fever by selecting higher body temperatures. Figure 9.7 illustrates how a lizard regulates its body temperature by volitional behavior (Purves *et al.*, 1992).

Although thermoregulation is the primary means for temperature regulation of ectotherms, it is also used to some degree by endotherms. When the option is available, most animals select thermal microenvironments that are best for them. They may change their posture, orient to the sun, move between sun and shade, and move between still air and moving air. More complex thermoregulatory behaviors are nest construction and social behavior such as huddling (Purves *et al.*, 1992).

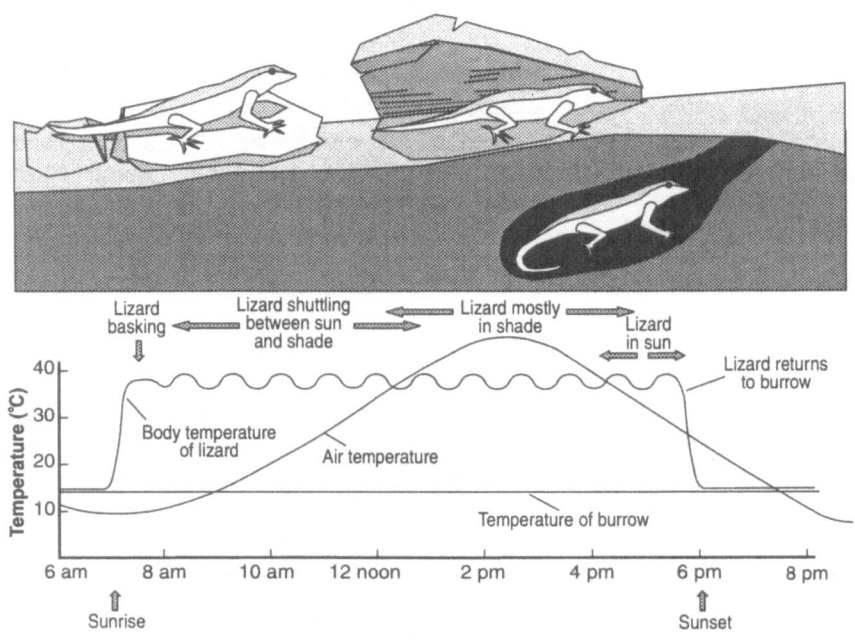

Figure 9.7. Volitional thermal behavior.

9.8.6. Quantification of Volitional Heat Behaviors

The volitional heat behaviors described above are specific to both the species and the environment. To quantify the amount of energy used in these behaviors, the amount of total energy utilized in the muscle activities associated with moving the body to obtain heat energy from the environment can be measured. Different species will use different amounts of muscle energy for a given thermal energy environment. An individual of a particular species will use different amounts of muscle energy as a function of changing thermal energy environments. However, for a specific individual in a specified thermal environment, the amount of muscle energy used to maintain a given body temperature can be measured. Due to the direct relationship between muscle contraction and information, the amount of motoneuron information that causes the muscle contraction behavior can be determined. The equations for behavior can be developed and, along with the measured muscle energy and motoneuron information, can be used to quantify the volitional heat behaviors in terms of the fundamental determinants of this behavior.

9.9. Summary

Volitional behaviors are defined and differentiated from autonomous and nonvolitional behaviors. Volitional behaviors are defined as those caused by information that has been adjusted by a central nervous system, whereas autonomous behaviors are primarily determined by genetic information and nonvolitional behaviors are mainly determined by chemical information and information in the nervous system that is not adjusted. The most easily observable volitional behaviors are, in the more complex animals, the result of the learning capability of the central nervous system.

Concepts are synthesized for the most ubiquitous volitional behaviors of food acquisition and coping with the thermal environment. Both of these behaviors are necessary for animals to survive. These concepts are based on the quantitative behavior principles developed in the early part of this book. The concepts were developed so they could be tested against extant behavior data.

The extant data is used to test the concepts. This testing resulted in a high degree of confidence in the application of basic quantitative behavior principles to volitional behaviors.

CHAPTER 10

Total Behavior of Individuals

10.1. Introduction

The research reported in the previous chapters provides the basis for determining the total behavior of individuals. The elements of behavior have been identified and the methods for quantifying these behaviors has been established, thereby providing the principles and tools for quantifying the total behavior of individuals. The next step in the development of a quantitative theory of living systems is the establishment of those methods and procedures for combining the elements of behavior given in previous chapters into the total behavior of individuals. This chapter takes this next step.

Total behavior is treated in this chapter in terms of combining the elements of the fundamental determinants of quantitative behavior, namely structure, capacity to direct energy, available energy, and information. The treatment follows the methodology of treating simple, easily observed, ubiquitous behaviors, and then proceeding to greater levels of complexity.

10.2. Methodology and Concepts

The methodologies and procedures used throughout this book are used again here for combining the behaviors presented in earlier chapters into the total behavior of an animal. That is, the concepts for combining behaviors are synthesized into a form that can be tested, then extant data are used to test the validity of these concepts. The data used were obtained mainly from readily available, time-tested, and accepted sources, such as encyclopedias, current college biology textbooks, and handbooks such as the *Handbook of Chemistry and Physics*.

The primary concept is that the total behavior of an animal is a combination of the same fundamental determinants of quantitative behavior that are applicable to the elements of behavior treated in previous chapters—structure, capacity to direct energy, available energy, and information. The elements of behavior can, in some cases, operate independently of one another; in other cases, they may be dependent on other elements; and still in other cases, there may be various degrees of dependence on other elements of behavior. There are three types of behavioral information: genetic, chemical, and neural. Each type of information causes the utilization of different types of energy. Genetic information causes energy in the form of matter to be synthesized into living structures organized in specific ways. Chemical information, such as hormones and enzymes, causes changes in living structures and causes parts of these structures to utilize energies in a coordinated manner. Neural information causes primarily the conversion of chemical energies into mechanical and heat energies.

The methodology is to proceed from the simple to the complex and from the most easily observed to the more difficult to observe. Therefore, it starts with single-cell animals and ends with the human species, and moves from the easily observed motion behavior to the complex chemical and molecular behaviors.

10.3. Single-Cell Animals

Single-cell animals, those of that Protista Kingdom, have the lowest level of complexity. Amoebas are protists that have often been described as the simplest form of animal life imaginable. Where animal-specific data are required, the amoeba was used to the extent possible as this animal has been studied extensively, resulting in a robust data base. The concepts above were tested first in terms of the most readily observed behaviors, that is, those associated with motion, then with more complex behaviors that can only be observed with the aid of the electron microscope.

In general, amoebas have single-cell structures that have pseudopods (false feet), extensions of their constantly changing body mass. They have no defined shape. Their basic cell structure is that of the eukaryotes, or cells that have a nucleus. Amoebas are almost transparent and are too small to be seen with the naked eye. They do not have a mouth or anus. A food vacuole can form on any part of its surface. Each freshwater species, such as *Amoeba proteus*, has a contractile vacuole on its surface whereas some saltwater amoebas do not. The surface structure of amoebas is sufficiently permeable to gases that oxygen can readily diffuse into and carbon dioxide can diffuse out of their bodies. Their structure and organization provide them with a capability to behave in appropriate ways in response to mechanical, chemical, light, and heat energies in their environments. They move away from mechanical forces such as that caused by the touch of a needle. They move toward food particles but away from such chemicals as sodium chloride and cane sugar. They will move away from strong light but toward weak light. And they move away from hot water.

The internal structures of amoebas are built on the general plan of eukaryotic cells. They have a plasma membrane, cytoplasm, and ribosomes. They also have other structures such as organelles that possess distinctive shapes and functions. One or two membranes enclose many of these structures. These membranes control what and how much gets into and out of these structures. Some organelles make special products and others take one form of energy and convert it into a more useful form. The structures and organization of the conversion of energy from one form to another are described in detail in Chapter 4.

10.3.1. Motion and Locomotion Behaviors

An amoeba *proteus* (A. *proteus*), when viewed with a microscope, looks like a microscopic mass of animated jelly. Upon closer observation, A. *proteus* are seen to be moving about with finger-like structures called pseudopodia. These are slowly thrust out from the jelly-like mass, and the remainder of the body flows into them. There is an almost constant streaming of protoplasm in the body. These behaviors are not random but are related to environmental phenomena as described above.

The mechanical energies in these observed motion behaviors can be described and quantified in known physical terms. For example, the energies of motion can be determined from the velocity of the motion, the viscosity of the fluid the cell is moving in, and the cell's size, shape, buoyancy, mass, and any other physical phenomenon that influences kinematic energy determinations. These energies can be calculated using the measures and equations for the kinetic energies of materials.

The mechanical energies internal to the amoeba that cause the observed motions can also be defined and quantified in terms of the amoeba's structure and mechanical processes. The extension and retraction of an amoeba's pseudopods provide the mechanical forces that result in the observable motion behaviors. Extension of a pseudopod is a function of the amoeba's structure. Just beneath the cell's plasma membrane, the cytoplasm is much thicker than the cytoplasm in the core of the amoeba—the thicker cytoplasm is called plasmagel and the more liquid core cytoplasm is called plasmasol. To form a pseudopod, the plasmagel thins in a certain area of the cell, allowing a bulge to form. Just under the cell surface, in the plasmagel, there is a network of actin microfilaments that interact with myosin to squeeze plasmasol into the bulge, thereby forming a pseudopod. As the plasma gel continues to contract, cytoplasm flows in the direction of the pseudopod. Eventually the cytoplasm at the leading edge of the pseudopod converts to gel and the pseudopod stops forming. Therefore, the basis for amoeboid motion is the ability of the cytoplasm to cycle through sol and gel states and the ability of the microfilament network under the cell membrane to contract and cause the cytoplasmic streaming that pushes out a pseudopod. The energy associated with the extension of a pseudopod is a function of the pressure to cause the flow of cytoplasm against the resistance to pressure caused by the pseudopod wall. Retraction of a pseudopod is a function of the amoeba's structure which allows the cell wall of a particular pseudopod to decrease its area and to cause the cytoplasm in the pseudopod to flow into the cell. The energy associated with the retraction of a pseudopod is a function of the pressure caused by the cell wall in reducing its area and forcing protoplasm out of the pseudopod.

The mechanical energies associated with the extension and retraction of amoeba pseudopods must be available from some source such as the chemical energy stored in ATP. As demonstrated earlier, the amount of free energy stored in ATP can be determined. Of course, this available energy ultimately comes from the energy stored in the amoeba's food intake. The manner in which chemical energy is utilized for these mechanical energies is a function of the structure and organization of the amoeba.

The data presented above validate, for the single-cell amoeba, the principle that an animal's structure provides it with a capacity to direct energy and that the energy in the motion and locomotion behaviors is a direct function of this capacity to direct energy. The data also validate the concept that the behavior of the single-cell amoeba can be quantified in terms of energy used in the motion and locomotion behaviors.

The observed motion and locomotion behaviors of amoebas have direction in addition to magnitude (speed). Recall that amoebas move toward food and low light, and away from sharp mechanical pressure, certain chemicals, and heat. This behavior requires the amoeba's structure to generate behavioral information in response to these environmental conditions and phenomena, and to route this information in such a way that causes energy to be utilized to provide motion in specific directions—either to or away from a given environmental condition. These data validate, for the single-cell amoeba, the concept that information is a fundamental behavior parameter for motion behavior.

The data on the motion and locomotion behaviors of amoebas, as detailed above, were sufficient to validate the concept that the fundamental principles of a quantitative living systems science apply to the single-cell amoeba's motion.

10.3.2. Energy Extraction Behaviors

An amoeba behavior that is related to, but less observable than, motion and locomotion is the acquisition of food and its conversion to the chemical energy ATP. An A. *proteus* can be readily observed creeping up on the flagellate *Chilomonas paramecium*, surrounding it with its pseudopodia and engulfing it. Microscopic plants, Protozoa, and organic particles also serve as food for A. *proteus*. The food, and a little water, is taken into the cell through the formation of a food vacuole on the body of the amoeba. Digestive juices secreted by the protoplasm go into the food vacuole where the food is converted to material that is absorbed into the protoplasm. Undigested wastes are expelled from the vacuole. This behavior is the first step in the conversion of energy in food to the energy stored in ATP. The energy in this behavior is the mechanical process of bringing the food into the body and the chemical energies of the reactions for digesting food, that is, converting food into materials that can be absorbed in the protoplasm and that can be further processed chemically.

The formation of a food vacuole is a behavior of the membrane tissue of the outer surface of the cell. This structure is capable of identifying food and configuring itself to form the vacuole. It requires energy to reshape the membrane into a vacuole of the appropriate size to engulf the food and the water taken in with the food. The process for bringing materials into the cell is known as endocytosis. Endocytosis involves the following steps: (1) the cell surface folds to make a small pocket that is lined by the plasma membrane; (2) the folding increases until the pocket seals off, forming a vesicle whose contents are materials from the environment; (3) the vesicle, enclosed in membrane taken from the plasma membrane, separates from the cell surface and migrates to the interior; and (4) usually the vesicle fuses with a lysosome and its contents are digested. (Lysosomes transport digestive enzymes that accelerate the breakdown of proteins, polysaccharides, nucleic acids, and lipids.) From the quantitative living systems point of view, this structure provides the cell with its capacity to direct energies associated with the first stage of converting food energy into the energies for chemical behaviors.

Following this step for converting food energy are the processes of glycolysis and cell respiration. These processes have been adequately treated in Chapter 4 and do not need to be repeated here. The processes of glycolysis and cell respiration take the products of endocytosis and convert them into available chemical energy in the form of ATP.

10.3.3. Maintenance Behaviors

The behaviors of single-cell animals that allow them to exist over their lifetime are not as easily observed as the relatively rapid behaviors of motion and energy extraction. It is the introduction of this longer time dimension that increases the difficulty of observation—time-lapse observation and measurement are required. In addition, the mechanisms for the maintenance of the structures, materials, and processes that keep a cell alive over an extended period can only be observed through the use of instruments such as electron microscopes and X-ray diffraction devices. However, sufficient observations of these maintenance phenomena have been made over the last half-century such that the mechanisms of cell maintenance are understood.

Two primary observations are that materials, such as digestive juices, are used up in the metabolic processes and replaced, and that the structure is continually being changed, created, and used up. To illustrate, in the food-engulfing process described above, the formation of vesicles is a structural change, and the digestion process inside a vesicle uses up some digestive juices. After this digestive process, the byproducts are transported in a vesicle to the outer membrane where the vesicle fuses with this membrane and releases the byproducts to the environment—a process called exocytosis. Observations made possible by electron microscopes and X-ray diffraction devices resulted in our current understanding of the creation of structure and materials in cells, and contributed to the central dogma of molecular biology. This central dogma can be stated as follows: deoxyribonucleic acid (DNA) codes for the production of ribonucleic acid (RNA), RNA codes for the production of protein, and protein does not code for the production of protein, RNA, or DNA. In other words, the production of structure and the other materials of a cell (all are protein) starts with the genetic information in DNA.

Although the production of protein starts with the genetic information in DNA, the mechanisms and processes for the actual production of protein are long, involved, and complex. These mechanisms and processes require (1) a structure in which the processes can take place, for example, the nucleus or the cytoplasm of a cell, (2) chemicals from which proteins are created, (3) sufficient chemical energy to drive the processes of producing RNA and protein, and (4) chemical information, such as enzymes, to cause the chemical reactions involved in the production of RNA and protein. These extremely complex processes are simplified below to the lowest level of complexity necessary to test the validity of the quantitative living system principles developed in this book.

The starting place for the production of the structures and materials for the maintenance of a cell is the structure of the DNA molecule. This molecule contains the genetic information which causes the production of each specific protein of the cell. It also contains the information for the replication of its own structure. The process of replication provides insight into the generation of genetic information, that is, the replication of genetic information contained in a molecule of DNA. First, there is a cell structure which contains materials from which to generate a new molecule of DNA. The structures of both prokaryotes (cells without a nucleus) and eukaryotes (cells with a nucleus) have a capacity to direct the energies necessary to generate a new molecule of DNA. Replication of the double-stranded DNA molecule includes all the following steps: (1) unwinding the two parent strands, (2) providing a primer for the synthesis of a new strand, (3) elongation of each of the daughter strands, (4) filling in the gaps in one of the strands (called the lagging strand), (5) connecting the gap fillers to complete the lagging strand, and (6) editing the newly synthesized strands for accuracy of replication. Each of these steps requires one or more specific proteins, many of which are the chemical information-carrying enzymes. Figure 10.1 is a schematic of the DNA molecule replication process.

The unwinding of the double helix is caused by two related enzymes called helicases, one of which attaches to each of the parent DNA strands. For the unwinding to occur, there must be energy available for this behavior. Energy to separate the strands comes from the hydrolysis of ATP. The separate strands are prevented from folding back on themselves by the attachment of single-stranded DNA-binding proteins to each of the separated DNA strands. These binding proteins hold the single strands in a configuration that binds readily to DNA polymerase III, an information-carrying enzyme. A pair of DNA polymerase III molecules at the replication location causes elongation of the leading and lagging strands. However, the discontinuous production of the lagging strand results in a repeated need for a new primer to start the synthesis of the next fragment. The primer is a short single strand of RNA, rather than of DNA, which is formed by the enzyme primase, one of several polypeptides bound together in an aggregate called a primosome. DNA polymerase III extends the primer. DNA polymerase I later causes the RNA primer segments to be replaced with DNA segments. Finally, the enzyme DNA ligase causes each newly completed fragment to link to the completed portion of the lagging strand.

In addition to catalyzing elongation of the leading and lagging strands, the enzyme DNA polymerase III plays another crucial role in DNA replication—it checks the accuracy of its own work. After it adds a monomer to a strand, it tests the new base pair to see that it is complementary to the nucleotide in the template

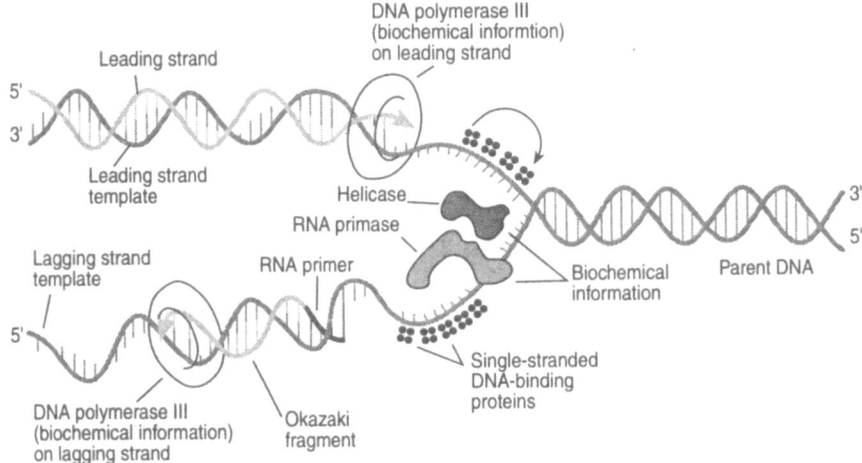

Figure 10.1. Schematic of DNA molecule replication process.

strand. If an incorrect nucleotide has been inserted, the DNA polymerase excises the erroneously selected nucleotide and tries again.

The synthesis of proteins involves the following: (1) the DNA molecule is partially unwound, as in DNA replication, (2) an RNA molecule is synthesized so as to contain the genetic information in the DNA molecule, that is, transcription of the DNA information onto RNA, and (3) genetic information in RNA and amino acids attached to specific RNA are used by ribosomes to synthesize protein.

Unwinding of the DNA molecule for RNA synthesis is somewhat similar to the unwinding process for the replication of DNA. However, for RNA synthesis, the DNA molecule is only partially unwound to expose the bases on its template strand that are to be transcribed. The DNA molecule strands are then rewound to return the molecule to its original configuration. In DNA replication, the DNA molecule is unwound, two new strands are synthesized, and each parent strand is rewound with a newly synthesized strand to form two new molecules of DNA. Synthesis of RNA requires the materials which will be in the RNA molecule, the energy in the form of ATP to drive the chemical reactions, genetic information from the DNA that is to be transcribed, and chemical information in the form of the enzyme RNA polymerase which causes the chemical reactions of the synthesis process. As the RNA is being synthesized, it peels away from the DNA template strand, thus allowing the DNA molecule to rewind. This transcription process produces three types of RNA: (1) messenger RNAs, or mRNAs, (2) transfer RNAs, or tRNAs, and (3) ribosomal RNAs, or rRNAs. All these RNA types are used in the production of protein.

The messenger RNA is so called because it travels from the site of its synthesis to the location where proteins are synthesized. In eukaryotes, it travels from the nucleus where it was synthesized to the cytoplasm where proteins are synthesized; that is, it carries the message. There is one messenger RNA molecule synthesized for each gene. The transfer RNAs recognize the genetic message and simultaneously carry specific amino acids. In this way, they translate the language of DNA into the language of proteins. The ribosomal RNAs become part of the organelle ribosome and constitute a major fraction of its composition.

The mRNA molecule forms as a complementary copy of one strand of the gene. It contains the information from that gene, so there are as many different messengers as there are different genes. The genetic information transcribed in an mRNA molecule can be thought of as a series of three-letter "words," that is, each sequence of three nucleotides (the "letters") along the chain specifies a particular amino acid. The "word" is called a codon. Thus, the mRNA molecule serves as a template on which the tRNA lines up to bring amino acids in the proper order into a growing polypeptide chain.

The tRNA molecules carry amino acids, associate with mRNA molecules, and interact with ribosomes. In effect, they bring the proper amino acids to the ribosome where the amino acids are combined to synthesize specific proteins. Each amino acid is combined with the correct tRNA by information contained on one of a family of activating enzymes known as aminoacyl-tRNA synthetases. Each activating enzyme is specific for one amino acid and for its appropriate tRNA. The enzyme reacts first with a molecule of amino acid and a molecule of ATP, producing a high-energy AMP–amino acid that remains bound to the enzyme—AMP is the chemical adenosine monophosphate. The high energy results from the breaking of the bonds in the ATP molecule. The enzyme then catalyzes a shifting of the amino acid from AMP to a terminal nucleotide of the tRNA, where it is held by a relatively high-energy bond. The activating enzyme then releases this charged tRNA, which provides the energy for the synthesis of a peptide bond that occurs during translation.

The structure of a ribosome, which provides the site for the synthesis of protein, consists of two subunits, a large, or heavy one, and a small, or light one. In eukaryotes, the large subunit consists of three different molecules of rRNA and about 45 different specific protein molecules, arranged in a precise pattern. The small subunit in eukaryotes consists of one rRNA molecule and 33 different ribosomal protein molecules.

Synthesis (production) of the protein used for developing the unit of genetic information is shown in Figure 10.2. The production of protein begins with the formation of an initiation complex, which consists of a ribosomal light subunit bound to the starting point of an mRNA chain and to a charged tRNA bearing the first amino acid. The heavy subunit of the ribosome then joins the complex. The polypeptide chain is manufactured based on "reading" the base sequence in mRNA at two ribosomal sites, A and P. When translation begins, the first tRNA is in place in the P site. At this time the second codon on the mRNA is exposed in the A site; a tRNA molecule pairs with the codon in the A site. The amino acid on the first tRNA bonds to the amino acid on the second tRNA. The first tRNA, having released its amino acid, dissociates from the ribosome. The ribosome then moves along the mRNA to the third codon, transferring the second tRNA and its two attached amino acids to the P site and exposing the third codon in the A site. A tRNA enters the A site, the two amino acids on the second tRNA in the P site are transferred to the amino acid on the third tRNA in the A site, and so on. The ribosome proceeds, "reading" mRNA codons and adding amino acids to the chain, until an "end chain" signal is reached on the mRNA. Then the completed polypeptide, final tRNA, and ribosomal subunits all dissociate from the mRNA and the process is complete. The protein resulting from this process can be used to maintain the cell.

In summary, the structure of single-cell animals includes cytoplasm, nucleic acids, and organelles (such as ribosomes) that provide them with a capacity to direct the energies necessary to manufacture proteins which can be used to maintain their structures and organization. The above data demonstrate that energy

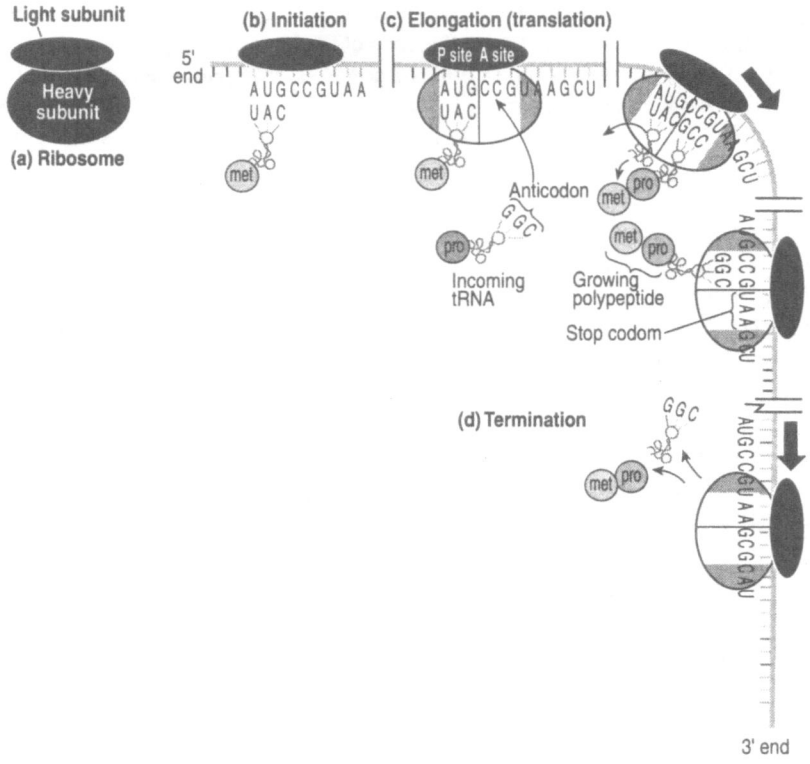

Figure 10.2. Production of the reference protein.

necessary to manufacture proteins must be available and comes from the ATP that results from respiration. The data also demonstrate that both genetic and chemical information are required to cause the behaviors that result in the production of protein.

10.3.4. Total Behavior

Animals do not exhibit the above behaviors in isolation but in a coordinated and complex manner. The behaviors are treated one at a time for clarity of validating the applicability of the living system principles to single-cell animals. The interrelationship of these behaviors is readily apparent. Motion and locomotion depend on the structure of the cell, which in turn, depends on the construction and maintenance of this structure through the manufacture of protein, which depends on the availability of chemical energy in the form of ATP, which depends on the conversion from energy bound up in food to the chemical energy stored in ATP, which depends on the motion of the animal to obtain food in the first place. Some of these behaviors take place simultaneously some of the time and not at other times. For example, the manufacture of protein can proceed without locomotion by the cell.

Combining the above behaviors at the detail level of behavioral energy, capacity to direct energy, available energy, and information described above would be a monumental, if not impossible, task. However, the total of these parameters can be obtained by other methods. The total behavioral energy can be obtained by measuring basal metabolism of the cell, metabolism under conditions of movement, and so forth. The total behavioral energies in motion can be obtained by kinematic measures. The energy that must be available for these behaviors can be calculated from the energy in the behaviors and the efficiencies of converting from input energy to the observable energies in the behaviors. Determining the number of quanta of information which causes total behavior is a more difficult task as each chemical reaction is caused by a quantum of information.

10.4. Simple Multicellular Animals

Simple multicellular animals are slightly more complex than the single-cell animals. The sponge is a representative simple multicellular animal of low complexity. Data on the sponge are used below for illustration and for further testing of the validity of the quantitative living system principles.

Sponges are so loosely organized structurally that they were difficult for early zoologists to classify as animals. Sponges are little more than loose aggregations of cells covered by an epithelium. Although the cells of sponges show some degree of differentiation, they are mostly variants of amoeboid types and each can perform a number of different functions. Furthermore, sponges have no mouth or differentiated digestive cavity; they are permeated by systems of canals through which a continuous water current is maintained. However simple, the sponge does have the structure to cause water to enter through innumerable pores scattered over the surface of its body, flow through a system of canals into a central cavity, the cloaca, and then out of a large opening, the osculum. This structure includes choanocytes or flagellated cells that line the canals of the sponge's body and propel water through the canals by their continual motion. It also includes a structure for controlling the amount of water allowed into the canals. The most obvious behaviors (Parker, 1918) of the sponge are the flow of water from its osculum and variations in this flow.

Flagella cells possess whiplike appendages, usually single or in pairs, which can propagate waves of bending from one end to the other in snakelike undulations. Flagella are built from specialized microtubules. There are nine fused pairs of microtubules, called doublets, forming a cylinder, and one pair of unfused microtubules running up the center. The motion of the flagella results from these microtubules sliding past one another, using energy from ATP to drive the process (Purves *et al.*, 1992). Flagella cells also possess the chemical mechanisms for initiating the continuous motion of the flagella. The structure of the flagella cells provides them with all the characteristics of an independent effector.

The early independent work of Wilson and Parker demonstrated that both the osculum and the dermal pores of the sponge are capable of opening and closing to vary the amount of water passing through the sponge. These structures also have within them the mechanisms for causing these openings and closings.

The cells constituting the sponge have metabolic behaviors common to the cells of all animals. That is, the cells have structures and organization which allow them to carry out the mechanical and chemical process of obtaining nutrients and oxygen from the environment and metabolizing them into the products and energy necessary for life.

Some sponges have structures which allow them to contract their flesh and thereby slightly change their form. The observation of this behavior can be traced back to Aristotle. In the fourteenth chapter of his fifth book on the history of animals, he makes the interesting statement that the sponge is supposed to possess sensation because it contracts if it perceives any movement to tear it up and it does the same when the winds and waves are so violent that they might loosen it from its attachment. This concept that the common flesh of the sponge is contractile is supported by Merejkowsky in 1878. He stated that if the sponge *Suberites* is partly out of water, it will curve its body until it is under water as much as possible, and if the body is then covered with water, it will return to its prior position. Parker (1918) states that much of the common flesh of *Stylotella* is contractile. To be contractile, the sponge structure must have contractile tissue and have the mechanisms that can cause contractions.

10.4.1. Motion Behaviors

An easily observable behavior of sponges is the motion of water flowing through the sponge. This is in contrast to most animals whose most easily observable behaviors are their motion and locomotion—sponges are not capable of locomotion. As mentioned earlier, water flowing through a sponge is caused by the motion of flagella, and the amount of water flow is also governed by opening and closing of the osculum and the dermal pores. The behavior of the flagella is treated first, followed by the opening and closing of the osculum and the dermal pores.

An elaboration of the flagella description given above provides more insight into the behavior of flagella. The central structure of a flagellum is called the axoneme. It contains a ring of nine pairs of microtubules. In the center of the ring may be one additional pair of microtubules, a single microtubule, or no microtubule. Microtubules are hollow tubes formed from polymerization of the globular polypeptide tubulin. Other proteins in the axoneme form spokes, sidearms, and cross-links. Sidearms composed of the protein dynein are responsible for generating force. Dynein is a mechanoenzyme that catalyzes the hydrolysis of ATP and uses the released energy to change its orientation, thereby generating mechanical force. When the dynein arms on one microtubule pair contact a neighboring microtubule pair and bind to it, ATP is broken down, and the resulting conformational changes in the dynein molecule cause the arms to point downward, toward the base of the axoneme. This action pushes the first microtubule pair upward in relation to its neighbor. The dynein arms then detach from the neighboring pair and reorient to their starting horizontal position. As the cycle is repeated, adjacent microtubule pairs try to "row" past each other with the dynein sidearms acting as "oars." But since the microtubules are anchored at the bottom, the axoneme bends, instead of elongating, as the microtubule pairs slide past one another. In ways not fully understood, the central microtubules and the other proteins that bind the axoneme together control the dynein action so all the microtubule pairs are not "rowing" at the same time.

The water flow behaviors just described can be expressed in terms of quantitative living systems principles. The energy in the water flow behavior can be quantified by hydrodynamic principles and procedures. That is, the measurable hydrodynamic parameters such as water volume, velocity, pressure, viscosity, and density can be used to calculate the energy associated with water flow. This is the amount of energy in the behavior. The structure and organization of the flagella and the water channels determine the capacity to direct energy of the system. The available energy for the behavior is provided to the flagella by ATP. The information which causes the flagella's mechanical motion is provided by the chemical information associated with the mechanoenzyme dynein. Since all the parameters of the behavior have been specified and can be measured or quantified, the complete behavioral relation can be expressed in terms of the fundamental equation and therefore the behavior can be quantified.

The osculum and the dermal pores influence water flow behavior by controlling the amount of water allowed into water channels by the dermal pores and the amount of water that is allowed to flow out of the sponge through the osculum. Both the dermal pores and the osculum control water flow in the sponge by opening or closing the water channels. The oscula of sponges are, in comparison with their dermal pores, relatively large openings and hence are more easily observed. In *Stylotella*, each finger ordinarily carries at its free end a single large osculum. The opening and closing of this osculum is the most obvious response of this sponge. The structure of the *Stylotella* osculum includes the oscular collar with a conspicuous sphincter whose contractile cells, or myocytes, are, in many cases, very close to the cavity of the osculum and therefore in direct contact with the water passing through it. The structure is relatively simple and does not contain neural tissue. It is this structure that provides the osculum its capacity to direct energies associated with the closing and opening behaviors of the osculum. These behaviors are accomplished by one set of contractile cells, in the nature of primitive muscle, working against another set of contractile cells or even against simple tissue elasticity. The primary information causing these behaviors is related to the strength of water currents in the environment. That is, the mechanical stimulation of a current of water causes the oscula of *Stylotella* to open or remain open. From a quantitative living systems perspective, the contractile cells generate information in response to mechanical pressures in the environment and use this information to cause contraction of the contractile cells. It has been previously demonstrated that energy must be available for contractile cell behaviors in the form of the ubiquitous chemical energy, ATP.

The opening and closing behaviors of dermal pores in *Stylotella* are somewhat different from those of the osculum. There are two structures and mechanisms for these behaviors in dermal pores. One is similar to the sphincter structures of oscula but the other has some differences. Like the osculum structure, dermal pores do not contain neural tissue. In addition, the environmental phenomena that result in the generation of the information which causes these behaviors are different for dermal pores. The dermal pores lead into a large subdermal space from which canals lead to the flagellated chambers. The structures of the dermal pores in some species have a membrane which can cover the pore, some have sphincter contractile tissue that can close the pore itself, and others have both these structures. Both types of structures have contractile tissue that can be activated to close and open the pore to the environment. That is, both structures have a capacity to direct energy that, when activated, results in the opening and closing behaviors. Environmental conditions that result in the generation of information that closes the dermal pores include ether, chloroform, strychnine, cocaine, and injury. Environmental conditions resulting in the generation of information that opens the dermal pores include atropine, weak cocaine, deoxygenated seawater, and warm seawater. Mechanical stimulation, except injury, does not result in the generation of information. Again, the available energy for these behaviors is in the form of ATP.

As already noted, sponges have motion behaviors that are observed as slight curvatures of their bodies under certain environmental conditions. These motion behaviors are made possible by the structure of the common flesh of sponges because it has contractile tissue. Although the structure of this common flesh has contractile tissue, it does not have nervous tissue. The information generated as a result of environmental conditions is through direct stimulation and not by nervous tissue. That is, the flesh has the characteristics of independent effectors.

10.4.2. Energy Extraction Behaviors

The structure of the sponge is, as previously described, little more than a loose aggregation of cells covered by an epithelium. As such, each cell is in contact with the environment of the sponge. The structure does not have a circulation system or an information system among the cells; therefore, the cells behave much as do single-cell animals. In energy extraction behaviors, the degree of coordination among the cells is essentially zero. Therefore, the energy extraction behaviors are, in essence, the same as those for the single-cell animals described above and need not be repeated here.

10.4.3. Maintenance Behaviors

The maintenance behaviors for the sponge are essentially the same as those for single-cell animals. These behaviors are treated in the section on single cell animals.

10.4.4. Total Behavior

The total behavior of the sponge is a combination of those behaviors caused by genetic, chemical, and neural information. At the level of organization of the sponge, the concept of neural information includes the stimulation of independent effectors even though there are no neurons. Even at this low level of organization, there are dependencies of one type of behavior on other types. For example, growth and replenishment of cells are functions of both chemical and genetic information. Also, contractile behaviors depend on both neural and chemical information.

The combination of these behaviors is adequate to determine the total behavior of an individual animal. It includes the development, physiology, and motion behaviors of these simple animals. The reproduction behaviors of these animals are not included in these total behaviors because sexual reproduction is a group behavior. Group behavior is the topic of a subsequent book which is to be based on the fundamental principles for individual behaviors developed herein.

Because the types of cells in a sponge are very limited and their metabolic processes are few compared with more complex animals, it is theoretically possible to identify the various chemical processes of metabolism and to determine the amount of energy associated with each chemical reaction in these metabolic processes. That is, it is theoretically possible to determine the amount of energy directed by each chemical reaction and then the total amount of energy directed by a cell at any instant of time. This total amount of energy is then a quantitative expression of the behavior of a specific cell. The total behavioral energy of the sponge is the summation of the behaviors of each cell.

10.5. Coelenterates

Coelenterates were selected for testing the quantitative behavioral concepts and principles as they apply to more complex animals than the sponge. The term "coelenterata" is applied to a large group or phylum of animals of a low grade of organization. However, their organization is of a much higher grade than the sponge. Coelenterata have tissue with definite form and individuality involving the differentiation of nervous and muscular systems and consequently efficient coordination of parts to cause movement and locomotion. Compared with the sluggish responses of the sponge, the movements of hydroids, coral animals, sea anemones, jellyfish, and other coelenterates are quick, though the movements of these animals are slow compared with those of vertebrates. This increased rate of response, compared with those of the sponge, is due to the fact that the coelenterates possess not only muscles but also nervous organs in the form of simple sensory surfaces which can cause muscle contraction more quickly than would be the case with direct stimulation. The sea anemone is used as an example because some of the earliest studies of receptor–effector systems were done on sea anemones, and there is a robust literature on this animal.

10.5.1. Structure and Organization

While the sponge is more like a loose aggregate of cells, the body of a coelenterate is based on a structural unit called a *gastrula*. A gastrula consists of a small sac with a single opening at one end (the *blastopore*). The wall of the sac possesses two layers of cells, one passing into the other at the margin of the blastopore. The inner layer is known as endoderm, the outer as ectoderm.

The majority of coelenterata are much more complex than the simple gastrula. They have bodies of almost infinite variety, but usually a prominent feature of the organization is the presence of a number of tentacles placed in a definite pattern around the mouth. The bodies of coelenterata are composed of a multilayered skin, a mouth, an esophagus, a digestive cavity, mucous glands, sensors, neural structures, muscles, and membrane support structures. The outer layer of the skin (the ectoderm) consists of membranous cellular tissue (epithelium) and nervous and muscular sublayers. The mouth is connected to an esophagus that leads to the digestive cavity.

Inside the digestive cavity, the endoderm is differentiated into regions. There are definite tracts of epithelium in which gland cells are concentrated; these, together with the adjacent endoderm, constitute the main digestive area. These tracts, the mesenterial filaments, can secrete a digestive juice and can themselves ingest food. However, in the most specialized cases the main absorptive region is not the filament itself but the endoderm on either side of it. The area of endoderm that will ingest food in such cases varies according to the size of the meal. The digestive enzymes of coelenterates are specialized for dealing with animal prey. There is a system for carrying nutrients from the digestive cavity to each cell in the sea anemone and for getting rid of the cell's metabolic byproducts. However, a circulation system in the true sense does not exist although currents are produced in the fluid contained in the coelenteron by means of endodermal flagella. The distribution of substances through the tissues takes place partly through the processes of amoeboid cells and partly by diffusion.

In addition to glands which make digestive juices, coelenterata have glands that produce (1) a mucous substance which will adhere to prey, and (2) poisons which can paralyze prey. These glands are located in the tentacles. The tentacles, in turn, have mechanisms for selectively secreting the mucous substances and for injecting poisons into prey.

The less active of the coelenterata, such as the sea anemone, have sensor cells that are affected by pressure, chemicals, and light. Early investigators discovered that pressure on the ectoderm from a blunt probe caused contraction of ectoderm muscles. They also discovered that the ectoderm and some of the endoderm were sensitive to pressure. and that various locations on the ectoderm of a sea anemone, such as *Metridium*, are more sensitive to pressure than others. These investigators also found that various chemicals could be sensed by certain cells of the sea anemone, as demonstrated by muscle contractions resulting from the application of these chemicals. The presence of light sensor cells was also demonstrated in the early part of this century by observing the reaction of the sea anemone to various light conditions: general application resulted in simultaneous moderate contractions of the main body, whereas directional light resulted in the turning of the oral disc toward the light.

Many coelenterata possess definite sense organs. These are found in the more active members of the group such as the *medusae*. There are two main types of sense organs. One type includes organs sensitive to light. These structures, though often simple, may attain the grade of eyes. The second type comprises organs of variable structure (*statocysts* and *tentaculocysts*) which include hard particles (*statoliths*) in their constitution. These latter organs sense gravity and are necessary for balance, as described in an earlier chapter.

The neural structure of coelenterata is very primitive. It possesses neither actual nerves nor any central controlling organ. Their nervous system's essential parts are the sensory cells and nerve cells of the ectoderm and endoderm, together with a network of fine fibrils connecting these cells with the muscle fibers and with each other. The nerve cells and fibrils constitute a "nerve-net" which runs in the deeper part of the epithelium. The net is sometimes fairly evenly distributed, sometimes better developed in one part than another, and sometimes concentrated in given regions. It is not a continuous network in the sense that the structure of one cell is actually fused with those of others. The cells are separate entities with their fibrils in contact, the junctions having the physiological properties of elementary synapses.

The muscle structure of coelenterata is also very primitive. It consists of a single layer or sheet of muscle fibers lying on the inner or outer surface of the mesogloea (the support for the other layers). The muscle sheets of coelenterates are normally only one fiber thick, and each fiber lies directly upon a supporting surface of mesogloea. Consequently, if strong localized muscles (as distinct from diffuse sheets) are required, these are generally attained by the simple expedient of pleating the surface of the mesogloea. The individual fibers are not independent structures but belong to and form part of epithelial cells of the ectoderm or endoderm. Therefore, the combination of each cell with its "muscle tail" (or tails) constitutes a musculo-epithelial cell, a structure characteristic of coelenterata but rare among higher animals. In more specialized cases the muscle fibers of the ectoderm become separate structures, the epithelial cells which produce them being of insignificant proportions.

The mesogloea support structures of coelenterata are highly variable. Sometimes they are gelatinous, sometimes almost cartilaginous, or again they may be very watery and unsubstantial. The other structures described above attach to and are held in place by the support structures. The type of and configuration of the support structure are functions of as well as caused by the genetic endowment of the particular species.

The combination of the structures described above make up the complete structure of a given individual. This complete structure determines an individual's capacity to direct energy and as such is a fundamental determinant of an individual's behavior.

10.5.2. Motion Behaviors

As discussed above, the structure and organization of coelenterata provide them with a much greater repertoire of motion behaviors than those of sponges. However, in accord with the behavioral concepts and principles elucidated herein, these more complex behaviors manifest themselves only when there is information to cause these behaviors. The coelenterata motion behaviors are treated below in the context of quantitative living systems principles.

The most readily observable and simplest motion behavior of the more passive coelenterata, such as the sea anemone, is capturing food through the motions of their tentacles. A less readily observable motion is the slower and less frequent behavior associated with body motions. A still less observable motion is the slow creeping of the sea anemone in moving from one location to another. More readily observable but more complex motion behaviors are those of free-swimming coelenterata, such as the jellyfish.

Most coelenterata capture food by sensing via chemosensors that food is nearby, seizing this prey with one or more of the tentacles, conveying it to the mouth with the tentacles, and swallowing it. The structures of the tentacles that make this behavior possible are (1) chemosensors on the tentacles that generate neural information from the chemicals in the environment that have been given off by the prey, and (2) a neural transmission system that provides information to the muscles.

Another method of feeding which may supplement or replace the typical one is used with small food particles. When these particles come in contact with some parts of a coelenterate, they become entangled in slime secreted by that part and then are transported by means of cilia acting in specific directions to the mouth. This method of feeding is probably the predominant one in certain sea anemones and jellyfishes.

The typical food capturing behavior of a sea anemone depends on differentiated sensors, neural nets, and muscle structures for its capacity to direct energy. The other feeding method given above depends on cilia and chemical information for its capacity to direct energy, much as the sponge depends on flagella and chemical information (the behaviors of flagella and cilia are similar). It is theoretically possible to measure the amount of energy in the muscle contractions for typical food capturing behaviors and the amount of information associated with these behaviors. Actual measurement data do not appear to be readily available. Again, there must be available energy in the form of ATP for these behaviors to occur. Data are not readily available on these measurements of muscle contraction energies and amounts of information that cause these contractions. It is also theoretically possible to quantify the energy in the behaviors of cilia, the chemical information associated with these behaviors, and the available energy, but again, data are not readily available on the measurement of these parameters.

All the readily observable motion behaviors, including the feeding behaviors just described, are the result of muscle contractions. These muscles can be identified rather easily. For example, early workers (Parker, 1918) identified at least 13 sets of muscles in the sea anemone *Metridium*. It is possible with current technology to measure the capacity to direct energy of these muscles. It is also possible to determine the neural information which causes the contractions of these muscles. Therefore, it is possible to quantitatively determine the behavior of these muscle behaving systems. The information to cause the motion behaviors of coelenterata is generated by distributed chemical, mechanochemical, and thermal sensors. Information can also be generated by definite sense organs, such as light and gravity sensors.

10.5.3. Metabolic Behaviors

Metabolism is the sum of all the chemical reactions taking place in an organism. Therefore, metabolism includes the energy extraction and the maintenance behaviors described above for the single-cell and simple multiple-cell animals. It was appropriate to treat these simple animals in terms of energy extraction and maintenance behaviors to obtain a detailed understanding of these processes. However, it is easier to treat the more complex animals in terms of metabolism due to the large number of chemical reactions in these animals, the relative ease of measuring metabolic rates, and the robust data base on metabolic rates.

Some obvious coelenterata metabolic behaviors include (1) the generation of digestive juices that operate outside the cells, that is, in the digestive cavity, (2) the generation of mucous substances, and (3) the generation of poisons. The digestive juices are synthesized from chemicals obtained from the environment. These environmental chemicals are largely in the form of the complex biochemicals that constitute the coelenterate's prey. A number of chemical reactions occur between the decomposition of environmental biochemicals and the end chemical products of the synthesis process. Energy is involved in each of these reactions— some reactions liberate energy while others require the addition of energy. Enzymes (catalysts) are also involved in each reaction. Indeed, the enzymes cause the reactions to occur. Recall that, in the principles of behavior developed herein, behavioral information causes energy to be utilized by a living system. There, as previously described, enzymes are chemical information carriers. The digestive juices may contain a number of chemicals, including specific enzymes. For example, the digestive juices of coelenterata include enzymes which cause the chemical actions associated with the digestion of the biochemicals contained in the prey of coelenterata. The synthesis of these enzymes is, from our conceptual viewpoint, the generation of information. Therefore, enzyme synthesizers are chemical information generators. As a comparison, in more highly organized animals, hormone glands are information generators.

The synthesis of mucus and poison follows essentially the same process as described above for digestive juices. From our point of view, it is the utilization of energy in the chemical reactions of synthesis and the chemical information that cause the reaction that is important in testing the validity of the behavioral principles and concepts. Again, the structure of the animal provides a capacity to direct energy (i.e., the mechanisms that support a chemical reaction), and it is the chemical information (i.e., enzymes) that causes the animal's capacity to direct energy to be utilized.

Another obvious behavior involving chemical reactions and information is growth. The biomaterials from which coelenterata are formed have been chemically synthesized from materials in the environment. Synthesis of the biochemicals constituting coelenterata requires the utilization of energy for each of the chemical reactions involved in the synthesis process. As with other biochemical reactions, those associated with growth are caused by enzymes, or chemical information.

The less obvious behaviors involving chemical information are those metabolic processes that are carried out in the cell, as described for single-cell animals. The structures of coelenterata are much more complex than those of a single-cell animal or a sponge and therefore have a capacity for more and varied behaviors. This is particularly true at the cellular level due to the much greater number and variety of cells in the more complex animals. Although it is possible to quantify metabolism of more complex animals using the quantitative living systems parameters, it would be both tedious and time-consuming. For animals with a high degree of complexity, such quantification may be close to impossible using existing tools and methods. Fortunately, there are other physical phenomena and methods which can be used to determining the total metabolic rate of the sum of the individual chemical reactions.

One of these physical phenomena is that all chemical reactions are accompanied by a utilization of energy. Another is that biochemical reactions are not 100% efficient and release heat in a form that is not usable to an animal. For example, the efficiency of the respiration process in converting glucose to water and biochemical energy (ATP) is approximately 63%. Still another is that an animal's size and metabolic rate are a function of its volume and surface. For example, the larger the cell, the more chemical reactions it can carry out in a certain amount of time (the rate of metabolism of a cell is a function of its volume), but a cell's rate of exchange of nutrients and waste products—including heat—with its environment is a function of its surface area. Therefore, as a cell grows larger, there is an ever-increasing mismatch between metabolic demands for exchange and ability to service those demands. The practical size of a cell is limited by this mismatch. Multicellular animals solve this problem through specialized cells and structures that transport food, oxygen, and waste materials. However, there is still a limit to size due to the need to dissipate the heat generated by chemical reactions within the animal. This relationship between metabolic rate and size means that larger animals have lower metabolic rates than smaller ones. These phenomena led Rubner (1902) long ago to postulate a law that metabolism is proportional to the superficial area of an animal. For example, the average heat output for the normal human male between the age of 20 and 40 is 39.5 calories per square meter per hour. These phenomena provide a means for relating the measurable heat output of an animal to the sum total of the energies utilized in an animal's numerous internal biochemical reactions.

The total quantification of the metabolic process in terms of quantitative living systems behavior parameters requires that the information involved in the metabolic process be known. As a quantum of chemical information is required to cause each of the biochemical reactions of metabolism, the total chemical information required to cause the total measured metabolic rate is the number of chemical reactions at any given time.

The total metabolic behavior of coelenterata can be quantified by measuring the heat energy being released to the environment for a specific time period, determining the efficiency of the biochemical reactions of metabolism, and determining the number of reactions for the specific time period. The number of reactions in the given time period determines the amount of information used in the metabolic behavior.

10.5.4. Total Behavior

The total behavior of coelenterata is a combination of the motion, metabolic, and maintenance behaviors. The motion behaviors can be quantified using the individual behavior as described above or they can be measured directly by the amount of kinematic energies used in the motion behaviors. No matter which way the behaviors of coelenterata are quantified, the above dissertation verifies that these behaviors can be quantified in terms of the quantitative living systems behavior principles developed in earlier chapters.

10.6. Vertebrates

There is a major difference in complexity and behavior between vertebrates and the single-cell, simple multicellular, and coelenterata species described above. This increase in complexity includes (1) the emergence of true organs in the vertebrates, such as, sensors, a central nervous system, differentiated muscles, and a chemical information, or hormone system; (2) significant increases in size; and (3) about a 1,000 fold increase in DNA (for example a typical bacterium contains about 1/1000th the amount of DNA in a single human cell (Purves *et al.*, 1992). This increase in complexity makes it very difficult to determine the total behavior of a vertebrate using the method of combining the individual behaviors of each behaving element in these animals. Fortunately, other methods are available, such as measuring the metabolic rates animals and the kinematic energy in their motions. Metabolic rate is normally divided into two components. One is called basal metabolism, which is the total metabolism under specific environmental conditions and when the animal is at rest (a minimum of kinematic motion). The second is the total metabolic rate and includes basal metabolism and the component of metabolism that results from motion. Basal metabolism is approximately the sum of energies associated with the autonomous behaviors treated in Chapter 7. Recall that the autonomous behaviors were treated in Chapter 7 in terms of the quantitative living systems principles: the energy in an animal's behavior, an animal's capacity to direct energy, the energy available to drive the behavior, and the information necessary to cause this behavior.

The kinematic energy method includes the actual motion of an animal with respect to its environment and the motions of internal elements of an animal's body with respect to other elements of its body. An example is the motion of the smooth muscles of an animal's gut. These kinematic energies can be quantified using the well-known principles and methods of physics. These behaviors are, in essence, the nonvolitional and volitional behaviors treated in Chapters 8 and 9, respectively. Another method for quantifying kinetic energy is to measure or calculate the amount of metabolic energy used in these kinematic behaviors. This metabolic energy is the difference between the total metabolic rate when an animal is exhibiting these kinematic behaviors and its basal metabolism.

The total quantitative behavior of vertebrates at a given time is a combination of the genetic, metabolic, and the kinematic behaviors. This total quantitative behavior changes with time as a function of the amount of information at any instant of time. For example, there is a difference in total energy between the behaviors associated with basal metabolism when there is minimal information associated with the skeletal muscles and the total energy associated with maximum work behaviors when there is a great amount of information transmitted to the skeletal muscles.

10.7. Humans

Mammals have the most complex behaviors of all vertebrates, and humans have behaviors that are considered to be the most complex of all mammals. These human behaviors are possibly due to their highly differentiated structures and organizations which give them a large capacity to direct energy. Human structures include a number of sophisticated sensors and effectors and the most highly developed central nervous organ of all animals. The existence and organizations of these structures are a direct function of genetic information, that is, the amount of DNA in an animal's cells. As previously noted, a human's cell contains 1,000 times more DNA than a single-cell animal such as a typical bacterium. This equates to a 99.99% difference between the DNA of a typical bacterium and a human. If the DNA sequences of humans are compared nucleotide by nucleotide with some other animals, they differ by the following percentages: chimpanzees 1.7%, gorillas 1.8%, orangutans 3.3%, gibbons 4.3%, rhesus monkeys 7%, and lemurs 22.6% (Sagan and Druyan, 1992). These genetic differences and similarities are an indication of the complexity of humans as compared to other animals. As one would expect, the differences between human structures and our genetically closest animal, the chimpanzee, are small compared to other animals. The differences and similarities in complexity of animals' behaviors and structures have been treated in Chapters 7, 8, and 9 in terms of autonomous, nonvolitional, and volitional behaviors, respectively. These chapters develop methods for quantifying behaviors in terms of the fundamental parameters of living systems and also develop the equations for the relationships among these parameters. The methods for combining these separate behaviors are provided here.

The behavioral energies in the autonomous behaviors treated in Chapter 7 can be totaled using animal calorimetry methods. This is possible because the minimum autonomous behaviors do not perform work on the environment, and the energies entering the body from the environment are released into the environment in the form of heat. This heat energy can be measured by calorimetry methods and is an individual's basal metabolism. Recall that basal metabolism is measured when an individual is lying perfectly still, sufficiently long after the last meal so that no digestion is going on, and at a temperature ranging between 30° and 35°C. These conditions negate such factors as muscular activity, emotional stress, food intake, and external temperature and represent minimum metabolism and heat production.

Total metabolism in the body may be determined by direct or indirect methods of calorimetry. The direct method consists of placing the individual in a suitably constructed chamber and measuring the amount of heat evolved. By the indirect method, the heat given off is computed from the respiratory exchange. If we want to determine only the rate of energy metabolism, all that is necessary is a suitable apparatus by means of which the amount of oxygen consumed during a given interval may be accurately measured. By determining the consumption of oxygen, the elimination of carbon dioxide, and the excretion of nitrogen in the urine, the data necessary for calculating not only total heat production, but also the nature and amount of each of the substances metabolized, are obtained.

These measurement techniques have been known for a long time. Atwater and Rosa (1897) invented a very accurate calorimeter for use in experimental studies on human subjects in the 1890s. This apparatus, improved by Benedict just after the turn of the century, measures heat production and respiratory exchange simultaneously. Many improvements have been made since these early beginnings. Basal metabolism can now be measured clinically with relatively simple and portable instruments that measure the volume of oxygen consumed during a short interval, from which the consumption per minute may be calculated. No special chamber is required; the subject simply breathes in oxygen from a bell-type spirometer. The expired air passes through a tank containing soda lime, which absorbs the carbon dioxide. As the oxygen is used up, the spirometer falls, the drop being automatically recorded in terms of cubic centimeters. From the oxygen consumption the basal metabolic rate may be computed by consulting tables of data giving the proper corrections for temperature, barometric pressure, height, weight, age, and gender.

Readily observed nonvolitional and volitional behaviors are typically associated with the contraction of muscle tissue. These muscle contractions perform work in the environment or work against internal resistance in the body. In either case, there is an increase in the metabolic rate which can be measured in terms of the increased heat associated with these behaviors. An indirect method for calculating these behaviors consists of measuring the metabolic rate while doing work, and by using a typical value for the efficiency of converting chemical energy into mechanical energy. This efficiency is the energy equivalent of mechanical work accomplished divided by the total energy expended while accomplishing work. In man, this efficiency is approximately 25%.

A more direct way of obtaining this behavioral energy is to calculate the energy used in kinematic motion under specific load conditions. These mechanical behaviors of man range from large motions, such as running, to very small motions, such as vibrations of vocal cords to produce sounds just detectable by the most sensitive instruments. Wells and Luttgens (1976) document the scientific basis for human motion in their text on kinesiology, which describes the motions of the human body and its parts in terms of physical parameters such as mass, time, distance, and velocity. The laws of mechanics apply to the description and measurement of these observed mechanical behaviors, and human mechanical behavior can be described, measured, and expressed in terms of energy (Kelley, 1971). For example, Garrett and Reed (1970) developed a computer graphics method for quantifying the segmental kinetic energies as well as total body kinetic energy.

Unfortunately, the motor units involved in these behaviors have not been identified. Had the motor units been identified, the quanta of information causing these behaviors could be specified. However, the process used for determining the number of motor units for Rana pipiens sartorius muscle, as described in Chapter 4, can be used for any motor unit.

The methods and processes just described for determining the metabolism and kinematic behavioral energies of human individuals allow the quantification of these behaviors in terms of the quantitative living systems principles developed earlier. The intent in this chapter is not to provide an exhaustive treatment of an individual's total behavior, but to demonstrate that the methods and processes are applicable to readily observable and ubiquitous behaviors. These methods and processes can be validated to the degree necessary to establish a high degree of confidence in them.

10.8 Summary

Methods and processes are identified for quantifying the total behaviors of animals in terms of the quantitative living systems principles developed in earlier chapters. These methods and processes are shown to be applicable to the spectrum of animals from the simple single-cell animals up through humans. Sufficient data are presented to test these methods and processes and to obtain a high degree of confidence that the quantitative living system principles are applicable to the total behaviors of animals.

CHAPTER 11

Summary And Findings

11.1. Introduction

The validation phase of quantitative living systems principles was completed by testing these principles as they apply to the total behaviors of animals. It is appropriate to summarize the research presented in the previous chapters and to report the principal findings of this research.

The major thrusts of this research are that fundamental principles can be developed for a quantitative living systems science, and that these principles can be developed using the methods and processes that have been used so successfully for the extant quantitative sciences. The existing quantitative sciences are geometry, astronomy, mechanics, heat, and electromagnetism.

Investigation of the ways that the existing quantitative sciences emerged resulted in the identification of the methods and processes common to the emergence of each of these sciences. The common features of these methods and processes are that a new fundamental measure emerged prior to the development of the principles of the new quantitative science and that each new quantitative science used the principles of previously existing quantitative sciences. These features are shown graphically in Chapter 1 in Figure 1.1. The quantitative sciences and their new fundamental measures are, respectively: geometry-length; astronomy-time; mechanics-mass; heat-temperature; electromagnetism-charge. All these existing quantitative sciences are for nonliving systems. It was postulated that the methods and processes for the emergence of the quantitative nonliving systems sciences could be used as a model for the development of a quantitative living systems science.

In addition, an analysis was performed of the methods and processes by which the quantitative nonliving systems sciences emerged from ideas to fundamental principles and laws in terms of the degree of confidence, that is, verification, at various stages from idea to laws. This process is shown graphically in Chapter 6 in Figure 6.1. These methods and processes were also used as a model for the development of the quantitative living systems science principles.

11.2. Formulation of Ideas and Concepts

The model for the emergence of the quantitative sciences requires that the subject of the science be defined and the characteristics of the subject be identified. In addition, the model requires that the behaviors of the subject of the science be observed for various conditions and that quantitative relationships be established among the parameters that are associated with these behaviors. For example, in astronomy the motion of the heavenly bodies is the observed behavior, and the primary parameters are time and length. In mechanics, the behavior is the motion of nonliving systems (matter), and the parameters are the mass of matter, its velocity, acceleration, and mechanical energy. The subjects of a quantitative living system science are things that live and the ways they behave under various conditions.

The basic idea for the quantification of living systems' behaviors is that these behaviors can be observed and quantified in terms of the energy utilized in these behaviors. Given this premise, it was deduced from the conservation of energy principle that there had to be energy available that could be converted into the energies observed in the behaviors. It was further deduced that the living system had to have a capacity to direct energy for it to convert the available energy into the energies in the behaviors.

It was observed that although a living system had a capacity to direct energy and there was energy available, behavior occurred only when there was something to cause the conversion of available energy into behavioral energy. It was postulated that this something was behavioral information.

11.3. Hypotheses

Based on the above ideas and concepts, it was hypothesized that (1) the behaviors of living systems could be quantified in terms of the energies in these behaviors, (2) the fundamental parameters of living systems behaviors are behavioral energy, a system's capacity to direct energy, available energy, and information, and (3) a formal quantitative relationship exists among these fundamental parameters. A further hypothesis was that there are sufficient extant data to test these hypotheses and to verify them to the degree necessary for them to have the confidence level associated with fundamental principles and laws.

11.4. Initial Testing and Verification

Readily available extant data were used to test the validity of the hypothesis that behavior is observable and can be quantified in terms of the energies in these behaviors. These data include the facts that the observation of natural systems' behaviors can only be observed via the energies utilized in the behaviors (Rosen, 1978), and that the energies in these behaviors can be either measured or calculated using existing measurement techniques and units of measure.

It was straightforward to deduce that if there is energy utilized in a behavior, then there must be energy available to a living system which it can convert to the behavioral energies. This deduction followed directly from the conservation of energy principle. It was also straightforward to deduce that the principles of energy provided a means for measuring or calculating this available energy.

It appeared at first to be obvious that if a system took available energy and changed it to behavioral energy that it had a capacity to direct energy. However, data were not found that would readily validate this hypothesis. Extensive research was performed to verify this hypothesis. This research is documented in *A Measure of Knowledge* (Simms, 1971). Material from this book is presented in Chapter 3. The key finding from this research is that when energy takes the form of structure and organization (systems), it acquires the capacity to direct energy. The capacity to direct energy is, therefore, a direct function of a system's specific structure and organization, and it can be measured or calculated using extant methods and units of measure.

Verifying the hypothesis that information causes a living system to exhibit observable behavior was not possible using existing data. There was no clear concept of information and no measure or unit of measure for information. Research to establish a fundamental measure of information was undertaken.

11.5. Invention of a Fundamental Measure of Information

A literature search revealed that a measure of information with the characteristics of a fundamental measure did not exist; one had to be developed. The model for the evolution of the quantitative sciences provided a starting point. First, the information had to have the characteristics of a fundamental measure. That is, it had to be based on a self-evident truth, it had to be tied to an invariant physical phenomenon, and it had to be of a form that could be accepted by the perceived community of users.

It was a self-evident truth that a person could "inform" a body part, such as a finger, to move and it would respond. Also, someone could "inform" another person to behave in some way, such as to move, and that other person would do so. Therefore, this "information" caused muscle contractions that resulted in the observed behavior. But this truth did not provide an understanding of the nature of information.

In reviewing existing concepts of measure, it was observed that there are similarities between the natures of energy and information. The notion of energy is highly abstract. Energy is weightless and occupies no space, yet it exists in a large number of forms. Energy cannot be directly measured—we measure energy by its effects upon matter, which can be weighed and measured. The notion of information is also highly abstract. Information is weightless and occupies no space, yet it exists in a large number of forms, such as genetic, chemical, hormonal, neural, and in forms that can be transmitted between individuals. Because a way to measure information directly could not be found, it was postulated that, like energy, information could be measured by its effects. This effect is that information causes living matter to utilize energy. With this understanding of the nature of information, it was postulated that methods similar to those used to measure energy could be used as a model to invent a measure for information.

The method used for developing a measure for energy was to select common matter which was in abundance, whose characteristics were well known, and that had been used in many experiments. This was water. A specific mass of water with a specified purity and under specific environmental conditions was selected as the "standard" matter to be used in the measure. Under these conditions, the heat energy input to the specified water standard that resulted in a specific increase in the temperature of the water, namely, one degree centigrade, was the unit of measure and is the calorie.

Using the energy measurement model, a measure for information was developed. The first task was to select an elemental living system "standard" equivalent to the water standard for the measure of energy. Muscle contraction behavior was selected as the standard because it is an easily observed behavior. Within muscles, the motor unit was selected as the elemental behaving unit because one impulse of "information" always caused a contraction of the muscle fibers associated with the motorneuron through which the information impulse travels. As the actual standard, a motor unit of the paired sartorius muscle of Rana pipiens was selected because this muscle had been the subject of many previous experiments and its characteristics were, perhaps, the best known, and there were sufficient data on this muscle for use in the development of an information unit of measure. Environmental conditions on the standard living system element (motor unit of Rana pipiens sartorius) were specified which are equivalent to the environmental conditions for water in the measurement of energy. For the living system standard, these conditions included the temperature of the muscle, the load conditions on the muscle, muscle with sufficient chemical energy that fatigue was not a factor, and the electrochemical impulse in the motorneuron sufficient to exceed the threshold necessary to obtain muscle contraction. One electrochemical impulse in the motorneuron was defined as a quantum of information and the amount of energy this quantum of information caused to be utilized by the standard living system behaving element was identified. A quantum of neural information was defined as the amount of energy caused to be used by the standard living system under the specified conditions.

Invention of a unit of measure for information using the methods just described verified that this measure of information has validity equivalent to the measure for energy. However, there are other measures of energy, such as the erg, which is a unit based on mechanical work as opposed to the calorie described above, which is a unit based on heat. An equivalent unit for information could be developed based on the amount of chemical energy this information causes to be used in a biochemical reaction (behaviors) by a living organism. Chemical information transporters are the enzymes which cause all biochemical behaviors (reactions). Since the energy in these biochemical behaviors can be measured or calculated, the information of a one-enzyme molecule in causing a biochemical behavior can be identified and can be measured. A specific ubiquitous biochemical reaction was selected and a unit for chemical information established. A similar method was used to select a specific gene that causes the synthesis of a simple molecule of protoplasm and to develop a unit of measure for genetic information.

11.6. Fundamental Equations

Invention of a quantitative measure of information allowed the development of an equation (see Chapter 5) between a living system's behavior, its capacity to direct energy, available energy, and information. A brief summary of the development of the fundamental equation for the behavior of living systems follows.

Development of an equation for the behavior of living systems required determination of the relationship among the four fundamental parameters previously identified, i.e., behavioral energy, a system's capacity to direct energy, available energy, and information. These parameters were represented by symbols as follows:

b = behavioral energy
k = a system's capacity to direct energy
e_a = the energy available to the system
i = behavioral information

A brief description of the characteristics of these parameters follows:

1. Behavioral energy is associated with every observable behavior of a living system, can be measured, has been directed by the living system, and causes the observed behavior.

2. A living system's capacity to direct energy is a function of the system's structure and organization and is characterized by the system's capacity to convert available energy into the energy in a behavior.
3. Available energy is energy in a form that can be used by a living system for conversion into behavioral energy.
4. Behavioral information, when applied to the effectors of a living system, causes them to convert available energy into behavioral energy.

A relationship was established among these fundamental parameters based on the nature of systems and these parameters, and on observed behavioral phenomena. The nature of a system's behavior is that it can only be observed by way of the energy in the behavior and can only be measured by this energy. It is readily observed that the energy in a behavior is different from the energy available to the system from its environment. For example, the energies in a muscle contraction, which result in an observable system motion, is in the form of work and heat, but the energy input to the system is chemical energy in the form of food. Since energy is neither created nor destroyed, energy in some form had to be available to the system for it to be converted or otherwise used for the resulting energy in behaviors. It is also readily observable that behaviors of systems vary greatly even though they have the same type of energies available to them from their environment. From this observable phenomena, it was readily deduced that different systems have different capacities to convert available energy into behavioral energy. That is, a system has a specific capacity to direct energy which is its fundamental nature. It was further observed that, although a living system has a capacity to direct energy, it does not utilize this capacity until there is an input of information.

From these considerations, it is apparent that behavior is a function of a living system's capacity to direct energy, available energy, and information. An equation for this relationship is:

$$b = f(k, e_a, i)$$

Further considerations of behavioral phenomena resulted in the determination that the behavioral energy of a living system is directly proportional to the system's capacity to direct energy, the energy available to the system, and the system's behavioral information. This relationship can be expressed mathematically as:

$$b = f(ke_a i)$$

The functional form of this equation was further investigated using the elemental behaving system for neural information, which is the motor unit. This investigation resulted in a relationship between a motor unit's behavior and the fundamental parameters associated with this behavior. That is, the behavior of a motor unit is equal to its capacity to direct energy multiplied by behavioral information, provided that there is sufficient available energy that the motor unit is not fatigued. In equation form, this equality is:

$b = ki$,　　　given that e_a is greater than k for each motor unit contraction

The unit of neural information can be as follows: one quantum of information is equal to 18.5 microcalories of behavioral energy. To obtain a measure of any other motor unit's behavior (contraction), the specific capacity to direct energy of this motor unit must be known. This equation is the relationship for the elemental behaving system for neural information—the motor unit.

The behavior of muscles composed of many motor units can be measured or calculated using this fundamental equation and knowing the capacities to direct energy of the various motor units of the muscle and the information inputs. The method for determining the behaviors of multiple motor units was developed.

11.7. Living Systems Science Evolution

The development of a quantitative living systems science equivalent to the existing quantitative nonliving sciences should follow the same evolutionary steps as these existing sciences. The major steps in the evolution of the quantitative sciences were identified and a model was constructed using these steps. The model was constructed in a form that could be used to guide the development of a quantitative living systems science. The essence of this model is that quantitative sciences evolve from ideas to laws through a number of stages with increasing levels of confidence in the ideas until the fundamentals of the ideas are invariant, or almost so, and the level of confidence approaches certainty.

The models used in the development of the principles of a quantitative living science start with the existing state of the science. Therefore, a brief history of the evolution of biology was given in Chapter 6 as it applies to the verification of the quantitative living systems concepts. The fundamental principles that have evolved in biology are:

1. Living matter consists of protoplasm. This property of matter separates living from nonliving systems.
2. The smallest amount and organization of living matter that can exist and exhibit the properties of life is the cell. The basic unit of organization of living things is the cell and all organisms are composed of cells.
3. The processes of nutrition and of respiration are fundamentally the same for all living things.

4. All living things are the product of living things. This doctrine is essentially equivalent to the physics doctrines of the conservation of energy and mass.

5. The fundamental mode of reproduction of animals and of plants is essentially identical.

6. Classification systems have been developed for the identification of living systems and of their similarities and differences.

7. Mendel's ideas on the most basic patterns of inheritance in sexually reproducing organisms have become Mendel's laws.

8. The central dogma of molecular biology is the idea that DNA (deoxyribonucleic acid) codes for the production of RNA (ribonucleic acid), RNA codes for the production of protein, and protein does not code for the production of protein, RNA, or DNA.

9. The mechanisms of protein production are well understood.

10. The concept of genetic information is understood and accepted.

11. The concept of hormonal information is understood and accepted.

12. The energies associated with muscle contraction, a biochemical reaction, and genetic generation of protein are understood and can be measured and calculated.

13. Information is an essential parameter of living systems.

These principles provide a robust qualitative basis for the development of a quantitative living systems science. The model for the development of a quantitative science indicates that a new quantitative parameter is necessary for the emergence of a quantitative science from a qualitative one. Because this new quantitative parameter (information) has been developed, a quantitative living systems science could be developed upon the existing qualitative concepts and on the measure of information. The principles of a living systems science developed in early chapters were used along with the extant biology literature to test and validate these principles.

11.8. Testing and Validation

Following the models for the evolution of quantitative sciences, testing and validation were performed first on the most obvious and universal behaviors, and then on less obvious and universal behaviors to increase the confidence level in the fundamental principles. Autonomous behaviors, such as heart contractions and breathing, were tested first as they are both obvious and universal for many animals, including humans. Nonvolitional behaviors, such as stimuli–response behaviors, were tested next as they can be obvious but are not continuous behaviors like breathing. Volitional behaviors were then tested. These validation tests were performed using neural information and muscle effectors. A final test was performed from another perspective. Individual animals were considered from the point of view of their total behavior which included those behaviors caused by genetic, chemical, and neural information.

The procedure used to perform these validation tests was to develop a concept of the behavior under consideration using the fundamental principles of a quantitative living systems science and to test this concept using existing biological data. The purpose of the tests was to determine if the fundamental principles could be used to quantify known biological functions of individual animals and their processes. The principles were deemed validated when they could provide quantitative measures for these biological functions and processes.

The results of the validation tests were that the fundamental principles of a quantitative living systems science have a high degree of confidence for all the biological functions tested.

11.9. Major Findings

A major finding is that the behaviors of living systems are a function of information and that information can be measured or calculated in the same way as other fundamental measures. The nature of information is as follows:

- Like energy, it is an abstract concept.
- It is weightless and does not occupy space, just like energy.
- It can only be observed and quantified by the energy used in living systems' behaviors.
- It is transient and perishable.
- It is defined as the ability to cause work.
- It can be measured or calculated by the work it causes.

It was found that the parallels between energy and information are striking. Also, it was found that genetic, chemical, and neural information all have the same basic characteristics; all three types of information have their own unit of measure (genin, biocin, and neurin respectively), and all can be measured.

Another major finding was the existence of a direct functional relationship between a living systems behavior, its capacity to direct energy, available energy, and information. An equation expressing this relationship was developed.

It was determined through testing that the principles of quantitative living systems science have a high degree of confidence. The principles are of a form that can be tested by other investigators. The author considers the confidence level associated with the principles of quantitative living systems science to be high.

11.10. Other Findings and Applications

Prior to developing a unit of measure of neural information, a general concept was developed that behavior is a direct function of information. This concept was applied to the problem of providing protection to surface ships from anti-ship missiles. The concept was to identify any and all possible behaviors of a given ship that could be used to help negate an anti-ship missile (ASM). The concept included determining the individual probabilities of any particular behavior of negating an ASM and then determining the conditional probability of a particular behavior negating an ASM given other possible behaviors. These conditional probabilities were then combined in such a way that the best cumulative probability of negating an ASM was obtained. The next step was to identify the information that could cause these behaviors. A "decision device" (a tactical engine) was inserted between the information sources and the behaving elements in order to obtain the best probability of negating the ASM. A system has been developed to implement this concept. This system is currently deployed on ships. This application confirms the basic concept and principle that behavior is directly related to information and that information is the direct cause of behavior.

Unit of Neural Information

A.1. Introduction

The development in Chapter 7 of the relationship between the behaviors of the heart and information provided the foundation for another measurement unit for neural information. Although the neural information unit based on a frog's leg muscle was a good starting point, a better unit may be one based on humans and one associated with more energy than generated by a motor unit in a frog's muscle. The human heart was selected as the behaving system for an information unit because it has all the characteristics necessary for the development of an information unit and it utilizes considerable energy per beat. One quantum of neural information causes all the heart muscle fibers to contract, heart behavior is invariant under specified conditions for a quantum information input, heart behavior is the most consistent and constant of the behaviors caused by neural information, and there is a robust literature concerning the heart.

Establishing the heart as the "standard" living system for developing an information unit requires the precise specification of this system's characteristics. Recall that the motor unit of the frog's muscle had to be specified in detail for the development of the neural information unit in Chapter 4.

Establishing the heart as the standard for developing an alternative unit of neural information provides additional benefits. Other behaving muscle systems can be compared to this standard, just as the heat capacity of materials is compared to the heat capacity of water.

A.2. Information Associated with Heart Behavior

The relationship between heart behavior, the heart's capacity to direct energy, available energy, and information was established in Chapter 7. This relationship, plus the characteristics of the heart, were used to establish a unit of measure for neural information. The relationship for heart behavior is the same as our basic relationship for behavior, where the parameters of the heart are used in the general equation for behavior.

A.2.1. The Human Heart's Capacity to Direct Energy

Although the heart exhibits the most constant behavior of all body muscles, like other muscles, its capacity to direct energy varies with the load on the muscle. Recall that in the development of the unit of neural information based on the sartorius muscle of the frog, the capacity to direct energy is a function of the load on the muscle. (See Figure 4.1 in Chapter 4 to see how the capacity to direct energy varies as a function of the load on this skeletal muscle.) The heart muscle can pump approximately four times more blood during periods of extreme exertion than it can during periods of rest. Just as was done in Chapter 4 in developing a unit of behavioral information, the specific conditions of the heart must be specified for the development of an information unit based on the heart. The most consistent and readily measured behavioral energy directed by the heart is that during rest. Also, there is a wealth of data on the energy associated with basal metabolism, that is, metabolism at rest. In addition, the conditions of basal metabolism measurements are well established. Therefore, the capacity of the heart to direct energy under conditions of basal metabolism was used.

The capacity of the human heart to direct energy can be quantified based on the data given in Chapter 7. There, it was reported that the work done by a human heart at rest is approximately 100 gram-meters per beat, which is equivalent to approximately 0.23 gram-calories per beat. Also, the data in Chapter 7 illustrate that the heart is approximately 20% to 28% efficient in converting chemical energy into heat and work. The total energy directed by the heart is the sum of the work and heat generated during each heartbeat. If the work per heartbeat is 0.23 gram-calories and the efficiency is 23%, then the total energy per heartbeat is 0.23 gram-calories divided by 0.23, or 1 gram-calorie per beat.

A precise measure of the total energy per heartbeat could be obtained using a detailed specification of (1) the heart to be used as the "standard" in terms of size, weight, structure, and organization; (2) the conditions under which the measurements are made in terms of basal metabolism state, temperature, atmospheric pressure, etc.; and (3) the load on the cardiovascular system. However, as with any fundamental measure, the unit of measure is arbitrary and must be agreed upon by the users of the unit. Due to these factors and the asthetic appeal of the unit 1, the measurement unit for the human heart's capacity to direct energy is suggested to be 1 gram-calorie per beat. Also, it is probable that a heart can be found that, with the proper conditions, will utilize 1 gram-calorie of energy per beat. This heart and the specified conditions could then be used as the "standard" heart to be used for determining the unit of behavioral information.

A.2.2. Available Energy

The behavioral relationship holds only so long as there is available energy for a heart contraction that is equal to or greater than the energy a heart can direct under the specified conditions associated with the measurement unit. For the suggested unit, there must be 1 gram-calorie of chemical energy available per heartbeat.

A.2.3. Information

One quantum (impulse) of information from the heart's pacemaker will cause the heart muscle to contract and to utilize the total amount of energy the heart can direct under the specified conditions.

A.2.4. Behavioral Information Unit

From the behavior relationship, Equation 5.15 from Chapter 5, for a motor unit (the heart acts like a motor unit in that all the muscle fibers contract in response to a quantum of information input) and the data presented above, the following obtains:

behavior = (capacity to direct energy) × (information)

given that the available energy is equal to or greater than the heart's capacity to direct energy per beat

Numerically, the equation becomes:

1 gram-calorie = (1 gram-calorie per quantum) × (quantum)

or

one Quantum of information = 1 gram-calorie of energy

The behavioral information unit, in terms of behavioral energy, can be stated as follows:

one Quantum of behavioral information will cause the reference living system (the human heart) to utilize 1 gram calorie of energy.

The reference heart and the environment are specified as follows: the total energy utilized by the heart per beat is determined under those conditions specified for basal metabolism, there is sufficient available energy in the form of chemical energy, and the heart is normal and healthy.

There are now two units for a quantum of neural information, one based on the sartorius muscle of the frog Rana pipiens and the other based on the human heart. As both can be useful they are retained; however, a notational distinction must be made between them. Either q or qs is appropriate notation for the quantum based on the Rana pipiens sartorius muscle. First, this unit is very small (18.5 microcalories) and can be related to the small q or qs, and second, the s in qs can stand for the sartorius muscle. Either Q or a QH is appropriate notation for the quantum based on the human heart. First, this is a unit with a high value compared to the unit based on the frog sartorius muscle, and the H can stand for the heart muscle. Any of the notations provides the proper discriminators, however, the q and Q notations are simpler and are similar to the heat notation where c denotes the gram-calorie and C denotes the kilogram-calorie. Thus, it takes 1,000 calories to make a Calorie. Similarly, it takes 18.5 million quanta to make a Quanta.

A.3. Specific Capacity to Direct Energy

The concept of a muscle's specific capacity to direct energy was developed in Chapter 5, along with a method for determining the numerical value of a muscle's specific capacity to direct energy. There, the numerical value of a muscle's capacity to direct energy was based on the sartorius motor unit of Rana pipiens as the standard reference system. The standard heart, described above, can be used as the reference system, rather than the sartorius motor unit, for determining the specific capacity to direct energy of muscles. Because of the large numerical differences between the capacity to direct energy of the standard heart and the capacity to direct energy of the standard Rana pipiens sartorius motor unit, there will be a large numerical difference between the specific capacities to direct energy of muscle calculated using these two standards.

Using the heart as a standard reference, a motor unit's specific capacity to direct energy is defined as the ratio of its capacity to direct energy to the capacity to direct energy of the reference heart. Because the capacity to direct energy of the reference heart is one calorie per quantum, the specific capacity to direct energy of other motor units is the same as their capacity to direct energy, provided their capacity to direct energy is expressed in calories. Also, the capacity to direct energy of motor units must be measured under the same load, internal conditions of the muscle, and environmental conditions as the standard.

As already demonstrated, the capacity to direct energy is a fundamental property of all living systems. The above development of specific capacity to direct energy relates this property to the unit of neural information which is based on a heart reference, thus providing a means for comparing the fundamental properties of elemental behaving systems. For example, the fundamental properties of tubular, arthropod, mollusk, fish, amphibian, reptile, bird, and mammal hearts can be compared based on their specific capacity to direct energy. Likewise, the capacity to direct energy of other elemental behaving units, such as skeletal muscle motor units and smooth muscle motor units can be compared among themselves, with each other, and with heart muscle. It is possible, using the methods described above, to determine the specific capacity to direct energy of the behaving elements of animals and to construct tables of this fundamental property. These tables would be similar to the specific heat tables which give the fundamental heat property of matter.

A.4. Information Associated with Cardiovascular Behavior

The behavior of the cardiovascular system is usually considered over some period of time, such as the minute and the day, as opposed to individual heartbeats. In Chapter 7, the "normal" heart rates for the individuals of a number of species were given in terms of the number of quanta (heartbeats) per minute. If the specific capacity to direct energy of the cardiovascular system of these species and the information quanta per minute are known, then the amount of energy per minute in the cardiovascular system behaviors of each species can be determined. These data can be used to determine the cardiovascular system behaviors for other periods of time, such as the day or the year.

Also, the behavior of the heart is a function of the load on the heart, that is, the amount of work it is required to perform. Both increased information rate and load change the behavior of the heart compared to its resting condition. An increased information quanta rate causes more contractions per unit of time and therefore increases the amount of behavioral energy per unit of time. Increased load changes the heart's capacity to direct energy, which is a function of the load on the heart muscle. This change in the capacity to direct energy is due to the change in the length of the muscle fibers of the heart caused by the increase in volume of the heart cavity during increased exertion.

Changes away from the basal metabolism (rest) conditions under which the information unit is defined can cause nonlinear changes in the heart's behavior. Although the only behavior the heart muscle can have is a contraction in response to an information input, the heart's capacity to direct energy changes with increased information rate and with the load on the heart muscle. Both these characteristics are caused by the properties of the heart. For example, the heart information generator can only generate quanta of information at some maximum rate due to the organization and structure of the generator, and the muscle itself requires some recovery time before it can accept a new quantum of information. This relationship is exemplified by the data in Chapter 7, which show that although all mammalian hearts are built on the same plan, the heart of a small mouse weighs about 0.15g and generates quantum rates of 1,000 per minute whereas the heart of the largest whales weigh about 200 kilograms and have heart rate less than humans. The "normal" quantum rate for a man is estimated at 68 to 76 per minute and for a woman at 74 to 80 per minute. However, with heavy exertion, the heart can operate at a higher rate. The upper limit of heart rate is a function of the refractory period of the heart.

Hearts are structured in such a way that they have a maximum capacity to direct energy. They have muscles that have particular shapes, a maximum length to which they can be stretched, and a minimum length to which they can be contracted. These characteristics determine the amount of blood in a heart chamber before and after a contraction and thereby determine the volume of blood pumped during each contraction. As the work performed by the heart is a function of the amount of blood moved through a given distance and against a particular pressure, the capacity to direct energy of a particular heart is relatively constant over the operating range of the heart. There can be slight variations from this constant value, just as there are slight variations in the heat capacity of water over the liquid range of water. Instruments are available which can measure and determine a particular heart's capacity to direct energy. The capacity to direct energy is a property of a heart and can be used along with the unit of measure for a quantum of information to determine the specific capacity to direct energy of a particular heart.

A.5. Summary

Another unit of measure for behavioral information is derived based on the human heart as the "standard" behaving living system. This unit was developed because its size is more directly applicable to the information used in the total behavior of a human and is thereby more related to the major thrust of this research. This larger measure for behavioral information is related to the smaller unit described in Chapter 4 which is based on the sartorius motor unit of Rana pipiens. The concept of a living system's specific capacity to direct energy was developed so as to provide a method for typifying this fundamental characteristic of living systems.

The information associated with cardiovascular behaviors under conditions of basal metabolism was treated to provide examples of how the information unit and the concept of specific capacity to direct energy can be used.

References

Aboul Wafa, M.H. (1964). *General Principles in Metabolism.* dar Almaaref, Alexandria, Egypt.

Atwater and Rosa, Report of the Storrs Agr. Exp. Sta. (1897). As referenced in M. Bodansky (1938). *Introduction to Physiological Chemistry,* 4th ed. John Wiley & Sons, New York.

Barham, R.M., and F.J. Boersma (1975). *Orienting Responses in a Selection of Cognitive Task.* Rotterdam University Press, Rotterdam, Netherlands.

Bayliss, W.M. (1919). The nature of enzyme action. In *Encyclopedia Britanica,* 11th ed., s.v. "biochemistry."

Becker, G., and J.W. Hamalainen (1914). Untersuchung uber die kohlensaureabgabe bei gewerblicher Arbeit *Skand. Arch. Physiol.* 1915:31, 198.

Benedict, F.G., W.R., Miles P. Roth, and H.M. Smith (1919). Human vitality and efficiency under prolonged restricted diet. Carnegie Institution Publication No. 280, Washington, D.C.

Brazier, M.A.B. (1977). *Electrical Activity of the Nervous System,* 4th ed. Williams and Wilkins, Baltimore.

Carlson, F.D., and D. R. Wilkie. (1974). *Muscle Physiology,* Prentice-Hall, Inc., Englewood Cliffs, NJ.

Carlson, F.D., D., Hardy and D. R. Wilkie. (1963). Total energy production and phosphocreatine hydrolysis in the isotonic twitch. *J. Gen. Physiol.* 46:851–882.

Consolazio, C.F., R.E., Johnson and L.J., Pecora (1963). *Physiological Measurements of Metabolic Functions in Man.* McGraw-Hill, New York.

Cooper, R., J.W, Osselton and J.C. Shaw (1980). *EEG Technology,* 3d ed. Butterworth, Boston.

Cork, J.M. (1942). *Heat.* John Wiley & Sons, New York.

Denton, D.A., and J.P. Coghlan, eds. (1975). *International Symposium on Olfaction and Taste (Bol. R.).* Academic Press, New York.

Galambos, R. (1962). *Nerves and Muscles.* Anchor Books, Doubleday and Company, Garden City, New York.

Garrett, G.E., and W.S. Reed (1970) Computer graphics: simulation techniques and energy analysis. In *Biomechanics, Proceedings of the C.I.C. Symposium on Biomechanics* (J. M. Cooper, ed.). Indiana University, The Athletic Institute, October 19–20, Chicago.

Grinnell, A.D., and L. O. Trussell. (1983). Synaptic strength as a function of motor unit size in the normal frog sartorius. *J. Physiol.* 338:221–224.

Haldane, J.S. (1922). Respiration. *Sillman Memorial Lectures,* No. 14.. New Haven.

Hertel, E. (1906). Experimenteller Beitrag zur Denntnis der Pupillen verengerung auf Lichtreize. *Arch. Opth.* 65:106–134.

Howell, W.C., and I.L. Goldstein, (1971). *Engineering Psychology: Current Perspectives in Research.* Appleton-Century-Crofts, New York.

Jeppsson, P. (1969). *Studies on the Structure and Innervation of Taste Buds: An Experimental and Clinical Investigation.* Struves Boktryckeri AB, Gotenborg, Sweden.

Kelley, D.L. (1971). *Kinesiology: Fundamentals of Motion Description,* Prentice-Hall, Englewood Cliffs, N.J.

Kuhn, T.S. (1970). *The Structure of Scientific Revolutions*, 2nd ed. The University of Chicago Press, Chicago.

Lewis, T. (1911) The mechanism of heart beat. In *Encyclopaedia Britannica*, 11th ed.

Lucas, K., revised by Adrian, ed. (1917). *The Conduction of the Nervous Impulse*. Longmans, Green and Company, London, New York.

McCartney, W. (1968). *Olfaction and Odors: An Asphresiological Essay*. Springer Verlag, New York.

Miller, J.G. (1978). *Living Systems*. McGraw-Hill, New York.

Millikan, R.A. (1935). *Electrons (+ and −), Protons, Photons, Neutrons, and Cosmic Rays*. The University of Chicago Press, Chicago.

Nauta, W.J.H., and M. Feirtag (1986). *Fundamental Neuroanatomy*. W.H. Freeman and Company, New York.

Nelson, C.V., and D. B. Geselowitz, eds. (1976). *The Theoretical Basis of Electrocardiology*. Claredon Press, Oxford.

Newman, J.R. (1956), *The World of Mathematics: A Small Library of the Literature of Mathematics from A'h-mose the Scribe to Albert Einstein, Presented with Commentaries and Notes by James R. Newman*, vol. 1. Simon and Schuster, New York.

Parker, G.H. (1917). Actinian behavior. *Exp. Zool.* **22**:193–229.

Parker, G.H. (1918). *The Elementary Nervous System*. J.B. Lippincott Company, Philadelphia, London.

Petrie, W.M., (1951). Measures and weights. In *Encyclopaedia Britannica*, 11th ed.

Prutton, C. F., and S.H. Marion (1947). *Fundamental Principles of Physical Chemistry*. The Macmillan Company, New York.

Purves, W.K., G.H, Orians and H.C. Heller (1992). *Life: The Science of Biology*. Sinauer Associates, Inc., Sunderland, MA.

Richtmyer, F.K., and E.H. Kennard (1947). *Introduction to Modern Physics*. McGraw-Hill Book Company, Inc., New York.

Rosen, R. (1978). *Fundamentals of Measurement and Representation of Natural Systems*. Elsevier North-Holland, New York.

Rubner, M. (1902). *Die Gesetze des Energiever brauch bei der Ernahrung*. Franz Deutiger Verlag, Leipzig, Vienna.

Sagan, C. and A. Druyan. (1992). *Shadows of Forgotten Ancestors: A Search for Who We Are*. Random House, New York.

Saltin, B. and P. Gollnick, (1983). Skeletal muscle adaptability: Significance for metabolism and performance, *Handbook of Physiology*, Section 10, Chapter 19. American Physiological Society, Bethesda, MD.

Shannon, C., and W. Weaver. (1949). *The Mathematical Theory of Communications*. University of Illinois Press, Urbana.

Simms, J.R. (1971). *A Measure of Knowledge*. Philosophical Library, New York.

Simms, J.R. (1983). Quantification of behavior. *Behavioral Sci.* **28**:274–283.

Simms, J.R., (1990). The fundamental equations for the behaviors of living systems. *Proceedings of the Thirty-Fourth Annual Meeting of the International Society for the Systems Sciences*, Vol. 2. (Portland, Oregon, July 8–13) 799–805.

Simms, J.R. (1991). Measurement of the information associated with muscle behavior. *Behavioral Sci.* **36**:140–147.

Simms, J.R. (1993). Criteria for general theories of systems. *Proceedings of the Thirty-Seventh Annual Meeting of The International Society for the Systems Sciences* (Hawkesbury, Australia, July 5–9), 224–232.

Simms, J.R. (1996). Information: its nature, measurement, and measurement units. *Behavioral Sci.* **41**:89–103.

Smith, R.S., and W.K. Ovalle, Jr. (1973). Varieties of fast and slow extrafusal muscle fibers in amphibian hind limb muscles, *J. Anatomy* (London.) 116:1–24.

Sperry, D.G. (1981). Fiber type composition and postmetamorphic growth of anuran hind limb muscles. *J. Morphology* **170**:321–345.

Steinach, E. (1892). Untersuehungen zur vergleichenden Physiologie der Iris. *Arch. Ges. Physiol.* 52, (2), pp: 495–525.

Sutherland, E.W. (1992). In W. K. Purves, G. H. Orians, and H. C. Heller, *Life: The Science of Biology*, 3d ed. Sinauer Associates, Inc., Sunderland, MA.

Voit, E. (1901). *Veber die Groesse des Energie bedarfes der tier in Hungerzustaende. Sietschrift fuer Biologie.*

Watson, J.D., and F. Crick (1953). The molecular structure of nuclecid acids. *Nature* 171:737–738.

Wells, K.R. and K. Luttgens, (1976). *Kinesiology: Scientific Basis of Human Motion*, 6th ed. W.B. Saunders, Philadelphia.

Yao, Y.M., and J.N. Weakly (1986). Differences in transmitter release and number of nerve terminals per motoneuron between two frog muscles. *J. Neurosci*, 6:498–506.

Zimmerman, J. (1982). MEG gets inside your head. *Psychology Today* **16**(4):100.

Index